PROJECT
MERCURY

EUGEN REICHL

Schiffer Publishing Ltd

4880 Lower Valley Road • Atglen, PA 19310

Other Schiffer Books by the Author: *Project Gemini*, 978-0-7643-5070-2

Other Schiffer Books on Related Subjects: *NASA: Space Flight Research and Pioneering Developments* by Hans-Jürgen Becker, 978-0-7643-3879-3

Translated from the German by David Johnston.

Originally published as *Projekt Mercury*
© 2013 by Motorbuch Verlag, Stuttgart, Germany.

www.paul-pietsch-verlage.de

Designed by Molly Shields
Type set in Avenir LT Std/Univers LT 47 Condensed LT

ISBN: 978-0-7643-5069-6
Printed in China

Published by Schiffer Publishing, Ltd.
4880 Lower Valley Road
Atglen, PA 19310
Phone: (610) 593-1777; Fax: (610) 593-2002
E-mail: Info@schifferbooks.com

For our complete selection of fine books on this and related subjects, please visit our website at www.schifferbooks.com. You may also write for a free catalog.

This book may be purchased from the publisher. Please try your bookstore first.

We are always looking for people to write books on new and related subjects. If you have an idea for a book, please contact us at proposals@schifferbooks.com.

Schiffer Publishing's titles are available at special discounts for bulk purchases for sales promotions or premiums. Special editions, including personalized covers, corporate imprints, and excerpts can be created in large quantities for special needs. For more information, contact the publisher.

Night Sky, Bright Stars and Milky Way Galaxy © EpicStockMedia
Night Sky, Stars Moon and Galaxy © EpicStockMedia

CONTENTS

PROJECT MERCURY BEGAN IN PEENEMÜNDE

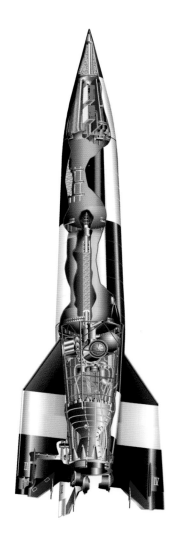

Project Mercury came to an end on June 12, 1963, just a month after astronaut Leroy Gordon Cooper returned to earth. Officially, it had begun four and a half years earlier on December 17, 1958. More precisely, its origins go back to October 3, 1942. That day saw the first successful launch of Aggregate 4, which had been developed in Peenemünde by a team under Wernher von Braun. Forty-six feet long, weighing fifteen tons and capable of speeds in excess of 3,107 miles per hour, it had a range of almost 185 miles and could reach a height of more than fifty miles, the edge of space. Aggregate 4's early history was equally unhappy and violent. Minister of Propaganda Joseph Goebbels gave it the name V-2, which stood for Vengeance Weapon 2. In all, 3,745 V-2 rockets were launched at targets on the European mainland and in Great Britain. Their warheads killed several thousand people, but even more lives were claimed by the inhuman conditions under which they were produced in Mittelwerk Dora, a mining tunnel in the Harz Mountains.

Then the war was over. As the Western Allies and the Soviets advanced into the heart of Germany, the fabled German large rockets were among the most sought after prizes. The victors took whatever they could get hold of: design drawings, parts, components and complete systems. Most sought after, however, were the men who had designed and built the rockets. The Americans initially had the advantage, reaching Mittelwerk Dora before the Russians. At about the same time, Wernher von Braun and his entire development team fell into their hands in Allgäu. He had been making preparations to hand over his team and all the plans since late autumn 1944. As planned, he and 125 of his associates were moved to the USA. Their extensive baggage included no less than 100 V-2 rockets. All that was left for the Soviets was a handful of managers of the development team. They did, however, capture the research facilities on the island of Usedom and

Literally all spaceflight around the world, whether in the USA, in Russia or in China, was based on the German original. The Aggregat 4 (V-2) was the first large rocket in the world and the first projectile capable of reaching the limits of space. The Redstone used in the Mercury program was a direct derivative of the Aggregat 4 and was created by the same design team.

THE DAWN OF SPACE TRAVEL

hundreds of executive-level technicians and engineers. They also discovered enough machine and rocket parts to initially resume production at Peenemünde. Later, people and materials were moved to the Soviet Union, where the V-2 was at first produced and developed by joint German-Russian teams. The knowhow transfer ended in 1951. The V-2, now called the R-1, began a second career in the Soviet rocket forces.

After the German technicians had given away their secrets and been sent back home, the Soviet Union immediately launched the development of intercontinental rockets. Unlike the Americans, at that time the USSR did not have a powerful strategic air force with which to transport its atomic warheads. For this reason they were quick to develop large liquid-fueled rockets for this purpose. The Americans, on the other hand, were convinced of the military value of their strategic bombers. In their opinion, the enormous weight of the early atomic warheads, five tons or more, ruled out the use of rockets from the outset.

A statement by the director of the Office of Scientific Research and Development, a man by the name of Vannemar Bush, offers an example of this thinking. At a congressional hearing in 1949, he

An Aggregat 4 is prepared for launch in New Mexico in the late 1940s.

declared: "There has been a great deal said about a 3,000-mile high-angle rocket. In my opinion such a thing is impossible and will be impossible for many years." These and similar statements caused American decision makers to adopt the view that intercontinental rockets would not become a subject of importance before the mid-1960s. And so, for many years, the V-2 rockets captured in Germany remained the largest rockets in the USA.

Of the 100 captured V-2s, sixty-six were launched at the US Army's White Sands Proving Grounds in New Mexico between April 1946 and October 1951. All were equipped with scientific instruments, and many had plants and animals on board. The telemetry data showed that the test animals withstood the ascent and weightlessness well. All died in the inevitable impact with the ground. There was no recovery system for them prior to 1951. The first photos of the earth from space were taken at this time. They gained far more attention than did the regrettable fates of the spiders, mice and rhesus monkeys.

With stocks of V-2 rockets rapidly dwindling, the USA was forced to find replacements. At the end of the 1940s, therefore, the Office of Naval Research ordered the development of two high-altitude research rockets, the Aerobee and the Viking. The Aerobee was capable of reaching altitudes of up to about eighty miles, while the large Viking was even capable of reaching 136 miles.

The USSR began exploring the upper atmosphere in 1947. Just like the Americans, they also began with their captured V-2s; but within just two years they had developed an improved version, the R-1, which not only served scientific purposes but was also used to equip special rocket forces. At the same time, they began development of large, multi-stage rockets. The story of the R-7 Semyorka, which would later put the first Sputnik into space and carried Yuri Gagarin aloft on his legendary orbital flight, began in 1951. The combatants did not know it, but the race for domination in space had begun, and the USSR had the better cards.

REDSTONE AND ATLAS

Since 1945, Wernher von Braun had been in the custody of the US Army at Fort Bliss near El Paso in New Mexico. The German scientists there had more or less been kept on ice, and for four years they did little more than launch the V-2s from captured stocks and work on several studies. But then the Korean War broke out in 1950. The army remembered its valuable specialists in the desert and in a cloak and dagger operation moved the just under 130-man group to Huntsville, Alabama, to the US Army's Redstone Arsenal.

Von Braun's mission there was to further develop the V-2 into a liquid-cooled battlefield weapon with

a range of 497 miles as quickly as possible. This rocket was given the designation Hermes C-1. Soon after the work began, the army changed its requirements profile. The rocket was now to have a range of only about 198 miles, but it was to be highly mobile, so that it could be deployed quickly to the battlefield. The rocket soon had little in common with the original design and in 1952 its name was changed. Not showing a great deal of imagination, they simply named it Redstone, after the place where it was born.

The performance figures for the Redstone were very similar to those of the V-2. Like its progenitor,

First flight of the Redstone on August 20, 1953.

its power plant operated with alcohol as fuel and liquid oxygen as oxidizer. It produced 33 kN (7,400 pounds) of thrust, was about sixty-eight-feet long and had a diameter of six feet. Its speed at burnout was 3,789 miles per hour.

The rocket's guidance system had an inertial platform and an analog computer. The machine's flight program was fed in on a punched tape, which was programmed prior to launch. This "computer" then issued guidance signals to the rocket at the intervals set on the tape. Flight control was by a combination of controllable aerodynamic fins and graphite control surfaces, which—like those of the V-2—functioned right in the thrust stream.

The first flight of the Redstone took place on August 20, 1953. At the time it was characterized as a "partial success," but today it would be called a failure. The rocket reached a height of just 19,685 feet. Soon, however, the rocket proved to be very dependable and the army nicknamed it "old reliable."

In 1955, a decision had to be made as to which rocket would be used to carry aloft the first satellite as part of the International Geophysical Year. The army proposed a development based on the Redstone and dubbed its plan Project Orbiter. This opportunity was wasted in wrangling between the two branches of the service. Under pressure from the air force, the army was excluded from further participation in Project Orbiter. Instead, the navy was given the task of designing a rocket based on the Viking as first stage and a modified Aerobee as second stage. This new rocket was called Vanguard.

The history of the Atlas rocket goes back even further than the Redstone. Initial studies began in 1946, under the direction of an engineer by the name of Karel Bossart of the Consolidated Vultee Aircraft Corporation (Convair). Based on this work, Convair received financing for the development of an Americanized V-2 with the project designation MX-774. Soon after completion of the first flight hardware, however, in 1947 the project was halted by the Truman administration and was not resumed until 1951.

In January 1951, by which time the Korean War had begun, the US Air Force gave a moderately

funded study contract to Convair for an intercontinental rocket under the project designation MX-1593. The rocket design that resulted was given the name Atlas by Karel Bossart and his team, but it would be four years until the plan was given the highest military priority. Given the immense technical hurdles, it was clear that it would be years before the rocket would be ready for use.

As feared, development of the Atlas turned out to be a huge collection of difficulties. The designers could fall back on earlier solutions or even past experience for almost none of them. The main problem areas were the rocket structure, the propulsion system, flight guidance and in particular reentry of the warhead into the earth's atmosphere. With respect to the structure, Bossart's people had a brilliant idea. They conceived the airframe as a balloon. While the structural material was stainless steel, it was thinner than paper. It was given the necessary stiffness by "inflating" it with helium. Structural weight was dramatically lower than previous rockets. Empty weight was less than two percent of its fuel weight. Despite this, it was capable of withstanding considerable structural loads. The air force selected the Rocketdyne Division

Karel Jan Bossart, the father of the Atlas. This photo was taken in February 1960.

The engine for the Atlas was derived from the launch booster of the Navaho (the white part of the twin-fuselage guided weapon).

of North American to develop the propulsion system. The company was able in part to fall back on existing equipment, for at the beginning of the 1950s the USA was developing the Navaho guided missile. This vehicle was the predecessor of the modern cruise missile, half aircraft and half rocket. The Navaho was a disaster from the beginning. The project was cancelled in 1957 after the then tremendous sum of 680-million dollars, equivalent to sixteen billion of today's dollars, had been spent on it. Just one important element survived the failed development program and underwent a reincarnation in the Atlas program: the power plant of the Navaho's takeoff booster, the XLR-83.

To avoid the then still difficult problem of sequential stage separation, Bossart adopted the principle of parallel stages. The Soviets had already used this method on their R-7 Semyorka. To this

day, it is still not known if a successful espionage operation led to Bossart's team selecting the same principle as the Soviets, or if it was the result of an independent technical analysis.

Parallel staging meant that all five of the Atlas' engines were ignited on the ground: the two booster engines (the ex-Navaho power plants), each of which produced just under 700 kN (157,000 lbs.) of thrust, the central main stage engine (the so-called "sustainer") with a thrust rating of 260 kN (58,450 lbs.) and the two guidance engines (the verniers), each with one Kilonewton (225 lbs.) of thrust. Some two minutes after liftoff, the booster rockets and their sliding frames were jettisoned. The sustainer took care of the remainder of the ascent.

While such a configuration was simpler in design, in payload ratio it was clearly inferior to two- or three-stage designs. On the other hand, the design's main advantage was obvious: avoidance of the delicate and still largely unresolved problems of stage separation and the ignition of liquid-fuel engines high in the stratosphere. All of the rocket's engines used a mixture of highly purified kerosene, designated RP-1, as fuel, with liquid oxygen as oxidizer.

Compared to the Redstone and the Thor, the Atlas looked uncommonly thickset and massive. With a length just under 75.5 feet, at its base it had a diameter of almost sixteen feet. The blunt nose cone further reinforced its brawny effect. Fully fueled, the rocket weighed just under 120 tons. Speed at engine burnout was 15,967 miles per hour. The initial versions had a range of about 6,213 miles, while later versions had operational ranges of 8,698 miles. Payload was 3,306 pounds. The basic idea behind the guidance system of the early Atlas versions was also very innovative, but in service it frequently proved problematic. The starting point of Karel Bossart's thinking was the fact that early computer systems were heavy, unreliable and consumed a great deal of power. After the Atlas had been designed as a one-and-a-half stage system, the fight for every kilogram of weight became vitally important. The solution was called radio-inertial guidance.

Sensors on board the rocket measured flight parameters such as aerodynamic forces acting on the rocket. This data was then transmitted to a computer on the ground. The computer constantly calculated the rocket's position, speed and direction and transmitted the necessary corrective inputs to the rocket's radio-inertial guidance system. The guidance system on board the rocket then swiveled the power plants as per the received data and the Atlas followed its flight path.

After all of its engines had burned out, the Atlas followed an unguided ballistic flight path. Soon afterwards the warhead separated from the rocket body and both flew towards the target separated by just a few hundred yards. The rocket burned up on entering the earth's atmosphere, while the warhead, which was fitted with a heat shield, entered undamaged. The advantage of this system was clear: the computer on the ground could be as heavy and complex as was needed and its weight did not affect the rocket. But the system's disadvantage also showed itself more than once in the history of the Atlas: if the radio link was lost, it automatically meant the end of the mission. The rocket would veer off course and have to be destroyed.

The very first launch of the Atlas took place on June 11, 1957. The Atlas A was equipped with just the two booster engines. The central sustainer engine was still missing. The mission was a failure, and the launch vehicle had to be destroyed at an altitude of 6,562 feet.

SPUTNIK SHOCK

In 1952, Albert Lombard, the defense department's scientific advisor, informed his superiors that the intelligence services had not discovered any signs that the Soviet Union was undertaking activities in the field of military or civilian earth satellites. At the end of 1953, the president of the Soviet Academy of Science announced that the Soviet Union would soon be in a position to send space vehicles into orbit and to the moon. This was dismissed as the usual grandiose Soviet propaganda.

In 1955, the US Navy began Project Vanguard. The goal of the project was to launch a series of small, unmanned satellites during the International Geophysical Year. Despite being called a "year," the International Geophysical Year was to last a total of eighteen months, from July 1, 1957 to December 31, 1958.

In August of the same year, the well-known Soviet physicist Leonid Sedov declared that the USSR was also considering sending a satellite into orbit. He continued on to say that this satellite would be heavier than the Vanguard satellite projected by the Americans. The news was ignored in the west. In 1956 a large preparatory conference for the Geophysical Year was held in Barcelona. The Soviets used the occasion to affirm their intention of launching an earth satellite to mark the event. In a 1957 press release, the Soviets even revealed the frequency at which satellites would be launched. On August 26, 1957, TASS, the Soviet news agency, reported that a new big rocket with intercontinental range had been tested successfully.

Many signs therefore pointed to the Soviet's firm intention of launching an earth satellite. Nevertheless, there was almost no official reaction to all these reports. The Americans merely shook their heads smugly over this, they thought, new example of typical Soviet boasting.

Only the report on the firing of the intercontinental rocket (which was nothing other than the first successful flight of the R-7 Semyorka) was acknowledged by the press with some murmuring. This culminated in the American press complaining that, "we obviously captured the wrong Germans." Scarcely anyone made the logical connection of a large operational payload-carrying rocket with a possible satellite launch. Instead, the press only looked into the military ramifications.

Only a few experts looked beyond the military horizon. They added one and one and came to the conclusion that the most probable date for the launching of a Soviet satellite would be September 17, 1957, the one-hundredth birthday of Konstantin Ziolkowski. The day came and went and nothing happened, and the last flurry of excitement about this topic went to sleep.

Then came the fourth of October 1957, and by the end of the day almost every man on earth could say at least one Russian word, and that was Sputnik. To say that the Americans were dismayed would be an incredible understatement. The Americans were also dumbstruck by the satellite's weight. Sputnik weighed more than 176 pounds, fifty-three times as much as the planned Vanguard satellite.

The furor had not died down when, just a month after Sputnik, a second Soviet satellite appeared in the sky. It had a dog on board and weighed more than half a ton. The humiliation of American science, research and military technology was complete.

Now panic broke out. A Senate committee under Lyndon Johnson was hurriedly formed to scrutinize national space flight activities. The American public looked hopefully to Cape Canaveral. And that was because of a bad mistake by White House press secretary James Hagerty. Desperate to tell the public at least something positive about the space program, he claimed that America was going to catch up to the Russians in a few days, for the Naval Research Laboratory was also going to launch a satellite using the Vanguard rocket.

The Vanguard team was completely dismayed by this announcement. The rocket was far from being ready to use. Official satellite launches were not expected until the spring of 1958. The first test with all three active stages was scheduled for early

December. In the unlikely event that it should work, they planned to place a tiny satellite the size of a grapefruit into orbit. To their great displeasure they now had to watch as scores of reporters arrived at Cape Canaveral to inform the public about America's heroic efforts to answer the Soviets. Nowhere was the test nature of the launch mentioned, and success was more or less felt to be a certainty.

The launch of the Vanguard was scheduled for December 6, 1957. A huge crowd of people gathered at observation points around Cape Canaveral and television brought the first major live coverage of a space flight event to millions of American households. It was the day when American national pride was to be restored. What followed is engraved deep in the national memory to this day. The rocket's power plant ignited and the Vanguard slowly rose about a yard from the launch pad. Then it fell back onto the pad and detonated in an enormous explosion before the eyes of the entire nation. America collectively sank into a state of shock.

Within hours the media began hounding the politicians. The government's earlier smugness and staidness were pilloried in giant headlines. The explosion of the Vanguard was an unparalleled disaster for the Eisenhower administration. This was all the more so, because in the meantime Lyndon Johnson's Senate committee had learned that, in 1956, the air force had been kicked out of von Braun's Redstone group because of jurisdictional wrangling.

Project Orbiter was immediately revived. Wernher von Braun promised to launch a satellite within just ninety days. That was only possible because he had correctly assessed the situation from the very start, disregarded the rules, and had a test rocket set aside from the Jupiter program. This version of the Redstone with the name Jupiter C was a special version of the basic model with lengthened tanks as used for reentry tests with the Jupiter warhead. Von Braun now equipped this rocket with a cluster of solid-fuel rockets, which he placed on its nose. They formed stages two to four. On January 31, 1958, just eighty-four days after von Braun had been given a free hand, the rocket put Explorer 1, America's first earth satellite, into orbit.

On March 17, the Vanguard was also launched successfully (after another failed launch on 15 February) and placed the three-pound Vanguard 1 into orbit. This was followed just nine days later by another Jupiter C with the seventeen-pound Explorer III (Explorer II had also been destroyed in a failed launch on March 5).

To the general public, it seemed that America's reputation had been at least partly restored. In any case they now had three satellites in orbit and the Soviets only two.

But then on May 15, 1958, the Soviets launched Sputnik 3 into space. With a weight of almost 3,306 pounds, it was fifty-six times heavier than the three American satellites combined. The Soviets seemed unbeatable. If they were capable of putting such a big satellite into orbit they were undoubtedly also in a position to put a man into orbit. This time, however, the Americans were not prepared to leave the field to the Soviets without a fight.

The Vanguard TV-3 (for Test Vehicle 3) exploded on the launch pad on December 6, 1957. The White House press secretary had previously given the public the impression that this was the first operational flight of the Vanguard rocket.

THE FOUNDING OF NASA

Although the first successful powered flight took place in the United States in 1903, technologically the Americans fell ever further behind the Europeans in the years after 1908. The great aviation theorists such as Ernst Mach, Ludwig Prandtl and Osborne Reynolds all worked in Europe. Record-setting aviators like Bleriot, Voisin, Fokker and many others also lived and worked there. To remedy this state of affairs, in 1915 the US Congress established an Advisory Committee for Aeronautics. Twelve members were named to the panel with the task of identifying problem areas in American aviation and overseeing scientific studies to solve these problems and put the results into practice. At its very first meeting the organization changed its name to National Advisory Committee for Aeronautics, or NACA for short.

In 1920, the NACA moved into its first research facility, the Langley Memorial Aeronautical Laboratory in Hampton, Virginia, named after the American aviation pioneer Samuel Pierpoint Langley (1834-1906). Over the years the NACA's tasks and responsibilities grew and the installation made a name for itself in aeronautical research. NACA was the address of first choice for young engineers and in 1939 the institution employed 523 persons, 278 of them in research.

The Second World War brought NACA another significant expansion of its responsibilities and tasks. During the war the Ames Laboratory near San Jose, California, and the Lewis Laboratory in Cleveland, Ohio (the present-day Glenn Research Center), were added to NACA's research facilities. By 1946 the organization had 6,800 employees.

The Korean War and the start of the Cold War arms race increased the demand for research personnel. Aircraft for speeds of Mach 1, Mach 2 and Mach 3 were designed and built at what today would be considered a breathtaking tempo. Huge bombers were created and altitudes of 12.5 miles and more were achieved. Towards the mid-1950s NACA was preparing to push to the limits of space with the X-15.

NACA viewed the work on these machines as a continuation of aviation with other means, for in the USA at that time work with rockets was viewed as an obscure science. This was attributable in particular to Robert Goddard who, though a gifted inventor, always conducted his research out of the public eye and jealously kept watch to ensure that none of his results reached the outside.

In addition to the solitary Goddard, there was just one other institution in the USA involved in rocket technology: the Guggenheim Aeronautical Laboratory. This installation was headed by respected physicist Theodore von Karman. The institute received its first government contract in 1939. This had nothing to do with space flight or even high-altitude research rockets, however, but instead for the development of a takeoff assist rocket for aircraft.

These takeoff-assist rockets were called JATO, which stood for jet-assisted takeoff. The word "rocket" was so linked to Goddard's theoretical research that its use for a production product was avoided at all costs. In 1944, the Guggenheim Laboratories were renamed, becoming the Jet Propulsion Laboratories. This—at the time intentionally misleading—title it still bears today. Until December 1958, however, the JPL was an independent research facility and not part of the NACA, which continued to have no particular interest in rockets.

This did not change until January 23, 1958, when Lyndon Johnson presented the senate committee's report on the state of American activities in the field of rocketry. Among the report's seventeen recommendations, one addressed the formation of an independent space flight agency. A variety of groups took up the proposal and it swiftly took shape. The military organizations were the only ones not especially pleased by the idea, for by then the three services had all included spaceflight programs in their planning and feared a loss of influence.

The navy's project was called MER I, an acronym for Manned Earth Reconnaissance. The space vehicle was extremely innovative looking. It was a cylinder with spherical ends. In orbit, which was to be reached by means of a two-stage booster, the structure would separate by means of two telescopic masts, after which a delta wing would be deployed and inflated.

The concept would have been too ambitious for its time. None of its components or subsystems had been tested, and everything, including the rocket that would propel it into space, would have to be developed from scratch. As a result, the project quickly disappeared into the curiosity cabinet of spaceflight history.

Much more promising was the US Army project Man Very High, which was later renamed Project Adam. To obtain research funding, the army characterized the program as a first step in developing new technologies for the rapid movement of special forces and tried to get the other two services on board. Project Adam did not foresee an orbital mission, but rather a suborbital flight at a height of 142 miles. The launch site was Cape Canaveral, with splashdown in the Atlantic off Bermuda. The approach was entirely realistic and was based on existing hardware and technology. The space vehicle was a modified Manhigh balloon capsule, which at the time was being developed and built by the air force for space medicine research and record attempts in the stratosphere. This capsule was 6.5 feet long and had a diameter of four feet.

The launch vehicle was to be a Jupiter medium-range rocket. For liftoff and reentry, the capsule would be protected by the heat-resistant nose cone of the Jupiter nuclear warhead. There was to be a passive passenger on board Adam with no influence on flight guidance. Adam was shelved in the summer of 1958 after it was criticized by the NACA and funding dried up.

The air force was pursuing two programs simultaneously at that time: the long-term Dyna Soar plan, an advanced single-seat space glider and the orbital crash program "Man in Space Soonest," a ballistic capsule. The air force was already well advanced, especially with the latter, which is not surprising given their claim that space was a logical extension of its previous area of responsibility.

The air force had reached an agreement with the NACA to carry out the "Man in Space Soonest" program. In May, however, before the cooperation agreement could be signed, President Eisenhower accepted Lyndon Johnson's proposal and submitted a bill to the Congress that would give the NACA responsibility for American space activities. In April 1958, the NACA, under its director Hugh Dryden, began organizing a national civilian space flight program.

Eisenhower instructed the Department of Defense to get together with the NACA and hive off the plans that should be carried out under the civilian roof. The parties quickly came to an agreement on almost every topic, with the exception of the manned program, which the air force very much wanted to retain. The talks began to falter.

By the beginning of July 1958, no decision had been reached about the organizational path of the manned program. Because of the uncertain situation, the air force had virtually stopped the flow of funding for Man in Space Soonest. This was unacceptable, however, because time was running out in the space race with the Soviets.

Both parties finally waited for a decision by the White House, and it came on August 18. On the advice of his scientific advisor James Killian, Eisenhower decided that, in future, the responsibility for carrying out civilian manned missions like Man in Space Soonest and its successors should be transferred to the civilian space agency. On the other hand the air force would retain the military-manned Project Dyna Soar.

Contrary to expectations, NACA chief Hugh Dryden was not named as the first administrator of NASA. Instead, somewhat surprisingly, the post went to Keith Glennan. This move was also made on Killian's advice. Dryden, who seemed too non-political to the White House, became Glennan's deputy and was responsible for all technical aspects of the new organization.

On September 25, Keith Glennan announced that the NACA would cease its activities on September 30, 1958. On the evening of that day, a Tuesday, 8,000 employees of the NACA, which had existed for forty-three years, left their place of work and Wednesday morning returned to their workshops, laboratories and offices as members of NASA.

On the left is NASA's Deputy Administrator Hugh Dryden. In the center is President Dwight D. Eisenhower, while on the right is NASA Administrator Keith Glennan.

THE HEAT BARRIER

Practically all of the engineers in the NACA, the aviation industry and the military shared the opinion that aircraft would gradually be developed to reach ever greater attitudes and higher speeds to ultimately become space vehicles. However, there was a problem that scotched that fine idea: frictional heat. The term "heat barrier" was born as an analogue to the sound barrier. Unlike the sound barrier, however, the thermal barrier seemed impenetrable. The problem with it was that one could not leave it behind after passing a boundary. Instead, with increased flight speeds in the atmosphere it grew exponentially and steadily increased.

For the X-15 it was anticipated that a temperature of 1,202 degrees Fahrenheit would be reached on the leading edges of the wing and the nose of the aircraft at a speed of 4,349 miles per hour. For a long time NASA looked for a material that could withstand this heat and ultimately found it in Inconel X, a nickel-chromium alloy.

But that was Mach 7. Mach 25 was a different matter. Convair engineers had calculated that, at a velocity of Mach 25 and a parabolic trajectory from an altitude of 932 miles to its target, the Atlas warhead would be subjected to temperatures of 3,632 degrees Fahrenheit on reentering the atmosphere. The warhead had to be shielded against this heat, but the only question was how.

In the first fifty years of aviation, aerodynamics had always led in just one direction: the reduction of drag in order to achieve higher speeds. As a result, the shape of aircraft had become ever more streamlined. The understanding that they could only progress in this one direction had become firmly anchored in the heads of the engineers.

When designing the Atlas reentry vehicle, Convair's engineers went back to the immense database that had accumulated over the decades. They then fed this information into their Univac and IBM computers in order, with their help, to determine the optimal shape for the reentry vehicle. The result was as expected: the machines calculated that the

best shape for this purpose was a long, needle-nose configuration, similar to that of the rocket-powered aircraft then under development.

However, when this configuration was tested in the wind tunnel of the NACA's Ames Center and during launchings of research rockets, it was discovered that this shape did not work. The vehicle was exposed to so much heat that the warhead was vaporized when it entered the earth's atmosphere. The seemingly logical approach of using a slender, aerodynamic vehicle so as to move as few air molecules as possible and thus produce as little heat as possible, failed in practice. The way out of this dilemma was discovered by Harry Julian Allen, an NACA aerodynamicist working in the Ames Laboratory. Allen later wrote of his discovery: "The people at Convair had simply switched off their computers too soon." He continued calculations from the place where they had stopped and came to the astounding result that the streamlined shape was the worst of all possible configurations. The shape had to be blunt, the blunter the better. Only by employing a blunt shape could the bulk of the heat be diverted around the object and discharged into the surrounding atmosphere. A wide resistor body would produce an arch-shaped shock wave in front of the vehicle. And this shock wave, the compressed air in front of the vehicle, would transport the greater part of the kinetic energy converted into heat during entry into the earth's atmosphere away from the reentry vehicle.

Even after Allen's groundbreaking discovery, however, the problem had not been entirely solved. The remaining heat produced by the vehicle resulted in temperatures that were still considerable. It made no sense to have a structure that was sufficiently heatproof if the payload inside the spacecraft was cooked.

Two different techniques were available to overcome this problem: the principle of the heat sink, and the principle of ablation cooling. Other techniques, for example active cooling using a film of liquid, were not considered because of weight

A photo from the year 1957. NACA aerodynamicist Harry Julian explains the concept of the blunt reentry vehicle.

or complexity considerations. The heat sink used efficient thermally conductive materials like copper, beryllium or molybdenum to absorb the heat, much as a sponge does with liquids. The mass of the cooling body had to be big enough to be able to absorb the introduced heat without melting the material. This method is effective but weight intensive, as a large, and in the case of copper heavy, quantity of metal is used to maintain an acceptable temperature. And there was the problem of how to quickly get rid of the metal body, which

had been heated to many hundreds of degrees during reentry, before the heat reached the interior of the vehicle. For this reason they turned to the second method, less out of conviction than bowing to necessity: ablation cooling. This saw a special surface coating on the outer skin transformed straight into a gaseous state by the frictional heat. The resulting gas stream carried the heat away along the shock wave.

At the end of 1957, however, the ablation technology had not yet been adopted. The first Atlas and Thor warheads still used copper heat sinks, but tests with the new technology were already being undertaken. One of the decisive tests took place on August 8, 1957, when a Jupiter C took an ablative Jupiter nose cone to a height of almost 621 miles. It then reentered the atmosphere 1,242 miles from launch point, at high speed and survived undamaged. But even after this test, the NACA still viewed the ablation technique with mistrust. The decision, which technique to use for the manned space program, was not made until two years later.

BIRTH OF THE SPACE CAPSULE

At the beginning of 1954, three development engineers at NACA Ames—Alfred Eggers, Julian Allen and Stanford Neice—wrote a study in which they examined the suitability of various aerodynamic shapes for vehicles reentering the earth's atmosphere. In their paper they reached the conclusion that a ballistic vehicle would be by far the simplest of the basic shapes that had been examined.

While it could not be controlled aerodynamically like other variants, its blunt shape—a cone or sphere—provided superior thermal protection thanks to a shock wave that formed in front of the vehicle. And it had by far the lowest weight. The three scientists' conclusion was clear: "The ballistic vehicle is a practical man-carrying machine, provided extreme care is exercised on entry into the atmosphere." First, however, they had to convince the NACA, which had traditionally focused on aerodynamic flight, of this view.

On October 15, a few days after the launch of Sputnik 1, the Ames Research Center began a conference dealing with the design of future manned spacecraft. A glider with delta wings and a flat undersurface was seen as the ideal solution. It was to become a reality sometime between 1963-1965.

Attending the conference in addition to Eggers and Allen was Maxime Faget, one of the future key personnel in the American spaceflight program. He took one thing in particular from the meeting: the development of a glider, no matter how beautiful and desirable it might be, would take too long to respond to the Soviets. Moreover, it would be years before a rocket would exist capable of carrying such a vehicle aloft. Contrary to their original intentions and their convictions as aerodynamicists,

he and many other engineers and scientists now began considering more intently how they could derive a machine from the ballistic reentry vehicles of the medium and long-range missiles in which they could put a man into low-earth orbit and bring him back again.

On March 10-12, 1958, the Air Force's Air Research and Development Command held a major conference in Los Angeles. More than eighty scientists from the air force, industry and NACA attended. Their task was to work out a scenario for the Man in Space Soonest project, in order to put a man into orbit and bring him back as quickly and as simply as possible.

And this was the plan: a ballistic vehicle would be used for the orbital flight and a parachute for landing. The weight of the vehicle was to be between 2,645 and 2,976 pounds. It was to have a maximum diameter of 6.50 feet and be at most eight feet tall.

NACA aerodynamicist Alfred Eggers in 1956.

The life support system was to be capable of keeping a man alive in orbit for forty-eight hours. And as it was not known at the time whether a man would even be capable of functioning under the conditions of spaceflight, all of the vehicle's systems were supposed to be designed to be fully automatic.

In this plan, the human on board was more a passenger than a pilot, even though he was supposed to attempt to carry out his tasks. The astronaut's seat in the cabin was to be arranged in the direction of acceleration and be rotatable. The rotatable seat was considered necessary because the vehicle was supposed to both take off and land with the nose forward. The maximum pressure load was anticipated to be 9 g, and the interior temperature during reentry was not supposed to exceed 149 degrees Fahrenheit. An ablative nose cone was supposed to provide thermal protection, while small retro-rockets were to slow the vehicle sufficiently to allow reentry into the earth's atmosphere.

One of the most difficult tasks was defining the physiological and psychological loads and

Maxime Faget, NASA chief designer of the Mercury capsule.

tolerances of the passenger. Lieutenant-Colonel John Stapp, already a legend for manning the rocket sled in tests at Holloman, was of the opinion that the first space passenger should complete both technical as well as medical training. And Major David Simons, known for his experimental flights in high-altitude balloons, declared that the astronaut should be continuously monitored medically and that constant radio contact with him was imperative.

Just four days after the air force meeting, a conference began in the Ames Laboratories concerning high-speed aerodynamics, in which an illustrious panel of people from the aviation industry, the military and NACA took part. Robert Gilruth, deputy director at Langley, instructed Maxime Faget to promote NACA's position with respect to the design of the future space vehicle. And by then this concept was clear: the ballistic capsule.

The configuration proposed by the Langley engineers in their presentation was a sugarloaf-like cone with an angular deviation of fifteen degrees from the vertical and a slightly bulged underside. It was ten feet tall with a diameter of 6.50 feet. For the role of heat shield they proposed a heat sink. The passenger would lie with his back to the heat shield during launch and recovery. Unlike the air force concept, Faget and his colleagues foresaw the entire space vehicle turning, positioning the heat shield in the direction of flight prior to reentry. The pilot thus did not have to change the position of his seat. The authors concluded with the words that, "as far as the reentry and recovery are concerned, the state of technical development is sufficiently advanced to begin the project of a manned satellite based on the shape of a ballistic reentry body." Although no one suspected it at the time, the basic specification of the Mercury space vehicle had been determined at these three meetings, even though the space vehicle's ultimate form had not yet been fixed. From then on other alternatives were no longer discussed. It was clear that the first American manned space vehicle—and as they then still hoped perhaps even in the world—would be a ballistic vehicle.

It is impossible to say from precisely where the term "capsule" for the manned ballistic space vehicle came. Its first use was in a report on the air force's planned five-year spaceflight program, which was submitted at the end of January 1958. This called for the development of reconnaissance, communications and weather satellites, as well as recoverable "data capsules" and manned "capsule test systems."

The air force subsequently used the term "capsule" to differentiate from the Dyna Soar glider concept. The name quickly took hold. By the late summer of 1958, the term "space capsule" was already in common use and from then on was also used in scientific publications.

Faget and his coworkers worked constantly to improve the design. The original configuration was a pointed cone with a slightly bulged heat shield on a broad underside. Wind tunnel tests at Langley, however, demonstrated that this shape was unstable in the subsonic range and that, most importantly, it was incapable of accommodating the parachute system.

After months of experiments, the team discovered a capsule design that combined the advantages of maximum stability, moderate heating and a sufficiently large parachute compartment. The shape of the heat shield had also been determined in general terms. It had a diameter of 6.50 feet, a curve radius of ten feet and a ratio between curve radius and diameter of five feet. In this form it diverted a maximum amount of heat away from the capsule during reentry.

One important question was the choice of a rescue system for the capsule passenger. The reliability level of the early ballistic missiles had been chronically bad. There was still little experience to draw on for the Atlas, as the pre-production versions of the A and B series were just then making their first test flights. Five of the nine flights up to July 1958 were failures, and trust in this launch vehicle was not particularly high. The logical conclusion was that the rescue system for the future space travelers must be

This was the initial concept for what became the Mercury capsule. It represented the state of development at the beginning of 1958.

extremely reliable. Air force planning for Man in Space Soonest had envisaged a so-called "pusher system," which had quite a few moving parts. This system would have seen solid fuel boosters on the base of the capsule fire to bring the space vehicle and its passenger to safety. Faget saw this design as too complicated and so he developed his own rescue system. His procedure was nowhere near as elegant as the air force design, but it was simple, fast and constructed from a minimum of components. It was a rocket mounted on a framework over the capsule that would pull the space vehicle behind it in event of danger. Combining its appearance and function, the system was given the name "escape tower."

The principle "quick and simple to realize" also applied to the launch vehicle. Again the air force had proposed a very complicated system, which would have required development of an upper stage for the Atlas. The NACA system, on the other hand, consisted of the Atlas without an upper stage. From a development point of view it was the simplest, but it had two major disadvantages. For one, the payload shrank to scarcely more than 1.3 tons, for another with this version a flight path had to be followed that could be critical for passengers.

A normal launch was not a problem. Things only became dangerous if a flight was aborted. Using the basic Atlas, a payload-optimized, steep ascent path had to be flown in order to place the capsule in orbit. If a flight was aborted, the descent angle would automatically be equally steep. That could mean that, under unfavorable circumstances, the occupant could be exposed to forces as high as 20 g, a possibly fatal value.

The air force had determined that the maximum g-force to which the occupant could be exposed without endangering his life was 12 g. Once again, however, Faget came up with the saving inspiration. He examined the couch that the air force intended to use. The passenger lay there—as it had always been—ON the couch. Faget asked himself whether the space traveler could not lie IN a seat that matched the shape of the body, so that the persons were supported from all sides and their tolerance for high g loads increased. And so he designed a light contour couch made of fiberglass. This couch was tested in the navy's big centrifuge in Johnsville. A volunteer, Marine Lieutenant Carter Collins, was able to endure loads of more than 20 g for six seconds in the couch. As a result, the basic Atlas was given the job.

CONTRACT FOR CAPSULE AND LAUNCH VEHICLE

By the beginning of October 1958, the preliminary specification for the space capsule for the Man in Space plan had been defined. The newly created Space Task Group of the equally new NASA sent this paper together with an invitation for an information event in Langley to a total of forty companies. The date for this bidder conference was set as November 7. Thirty-eight of the forty companies contacted sent representatives.

The briefing that day was given by Faget, and several of his colleagues; its purpose was to present the plan and the timetable to the companies and to gather ideas that could be incorporated into the ultimate specification. Nineteen of the thirty-eight companies subsequently declared an interest in taking part in the invitation to tender. These companies received a fifty-page document titled Specifications for Manned Space Capsule on November 14. The deadline for submission of proposals was December

11, 1958. Eleven offers were received by that date. They came from Avco, Convair, Lockheed, Martin, McDonnell, North American, Northrop, Republic, Douglas, Grumman and Chance-Vought. The next task was to select from the proposals.

After intensive consultations, two companies were left, which had received almost the same evaluation: Grumman and McDonnell. On January 12, 1959, NASA informed McDonnell that it had won the contract. The contract was signed just three weeks later. Its total value was $19,450,000, roughly 400 million in today's dollars. This figure seemed low for the development and construction of a manned space vehicle, and in fact it was. But as quickly as the contract came about, it was well handled by McDonnell and in the years that followed its value rose steadily. By the end of the year it was already worth $41,000,000, mainly due to the rapidly growing complexity of the capsule.

Faget and his colleagues had written the preliminary specification for the capsule with its fifteen subsystems in remarkable detail. The McDonnell engineers now began detailing these specifications down to the component level, filling gaps, producing design drawings for each component and preparing the workshops for the development and production of the capsule.

The first major milestone of the contract called for McDonnell to complete the detailed specification and build a full-size mockup including escape tower by March 17, 1959.

For the planned suborbital test flights of the capsule, the Redstone and the Jupiter were the only reliable boosters in the American arsenal. Both came from Wernher von Braun's renowned team. The Jupiter was a development of the Redstone and was produced by Chrysler. Von Braun's people guaranteed that both types could be made available within twelve to fourteen months. So, in January 1959, the Space Task Group went to the Army Ordnance Missile Command and concluded a delivery contract for eight Redstones and two Jupiters.

Just a month earlier, on December 8, 1958, a contract had been signed with the Air Force Ballistic Missile Division for the purchase of nine Atlas launch vehicles. This organization was the military customer for Convair's Atlas rocket. As required, the rockets were taken from Convair's production line and then modified for use by NASA. The effect on the launch vehicles was the same effect as on the capsule, and costs quickly got out of hand. Just two months after NASA had ordered the rockets for a unit price of 2.5-million dollars, the Air Force Missile Division came with the news that the rockets would now cost 3.3-million dollars each.

In May 1959, the Army Ordnance Missile Command also reported a price hike. The Redstones and Jupiters would now cost a total of eight-million dollars more than when ordered. The cost increase was attributed to the need for the army to share the cost of the development center in Huntsville. A Redstone thus cost 2.7-million dollars

NASA Administrator Keith Glennan (left) with Senator Lyndon B. Johnson, chairman of the Senate Committee for Space and Astronautics.

and a Jupiter 2.9-million. This was only slightly less than the price for the Atlas, which was ten times as powerful.

Keith Glennan subsequently complained to the Secretary of Defense, but it did no good. In order to at least remain halfway within the originally proposed cost framework, the NASA leadership decided to cancel the tests with the Jupiter C. This was only possible, however, because early on they had provided for an alternative to the expensive carrier.

LITTLE JOE
AND BIG JOE

For some time, thought had been given to replacing the expensive rocket test flights, at least in part, with aircraft and stratospheric balloons. In fact, in the summer of 1958 a test program had begun in which dummy capsules were dropped from an Air Force Lockheed C-130 Hercules transport to test the opening behavior of the parachute.

The use of stratospheric balloons promised to enable the capsules to be tested under "near space conditions." The use of balloons was prompted by the flights by David Simons as part of the Man-High project. They considered taking the capsules to a height of twenty-one miles and then executing the entire reentry sequence, beginning with the firing of the retrorockets to the opening of the drogue chute and the main parachute to recovery after the drop.

The Space Task Group had already issued contracts to the navy and air force in order to carry out manned stratospheric flights with the capsules, but then stepped back from the idea. Preparation for such a flight with training of the pilots and development of instrumentation would have taken more than a year and cost significantly more than the solution that Maxime Faget already had up his sleeve.

This solution envisaged the development of extremely simple and cost-effective rockets specifically for test purposes. The rockets were to be capable of testing and qualifying original size and weight capsules under all possible conditions in the first phase of flight. They were to have performance figures that came as close to those of the Redstone as possible. To achieve this they were to use standard army solid-fuel rockets, which would be bundled to provide the necessary performance. Existing components and systems were to be used wherever possible. Everything that was complicated or expensive was to be dispensed with.

A cluster of two or four modified Sergeant artillery rockets was to provide the core drive

The concept of the Little Joe also went back to Maxime Faget. Here the launch of mission LJ6 on October 4, 1959.

components. Two versions with slightly different power outputs would be used, either named Castor or Pollux. These clusters of two or four rockets were bolstered by four Recruit rockets, which formed an outer ring around the inner cluster and provided extra thrust for launch. When all eight engines were fired simultaneously on the ground, they provided a takeoff thrust of 1,700 Kilonewtons, or just over 382,000 pounds. Theoretically this was sufficient to deliver a space vehicle weighing 3,968 pounds on a ballistic trajectory to an altitude of ninety-nine miles. In practice, however, this value was never achieved.

The rockets' acceleration profile was supposed to simulate those of the Redstone and Atlas in the first seventy seconds of flight. To keep the price low and shorten development time, they were to have no expensive electronic components such as a flight guidance system and their ability to transmit telemetry was limited.

The cross-section of the central cluster of four rockets reminded the designers of a Hard Four in the dice game craps, a throw that is called Little Joe by players. This name gradually worked its way into the official documentation. Later, when the four Recruit rockets were added, the name solidified, and there were also the four large stabilizing fins at the base of the rocket structure. All of this contributed to the name Little Joe ultimately coming to be used for the entire rocket.

In the end the Little Joe cost just 200,000 dollars per rocket; it needed almost no infrastructure and could be launched from the NASA launch site on Wallops Island in Virginia. The call for tenders for detail development and construction followed that autumn. Twelve companies tendered submissions and in the end, on December 29, 1958, North American's Guided Weapons Division won a contract for six rockets, one test airframe and a mobile launch pad.

Little Joe was easily able to meet the criteria for the planned suborbital flights with the Redstone, but it was unsuitable for determining orbital test parameters. A kind of Big Joe rocket was needed for this, and the choice was obvious. Big Joe could only be the Atlas itself.

It was clear that they would not be able to afford many Big Joe flights. A single Atlas flight cost as much as fifteen Little Joe missions, therefore just two were planned. And they had to take place as early in the program as possible if they were to establish the space capsule's fundamental base criteria. In particular, the basic question of the heat shield remained open. Should it be a metal heat sink or an ablative shield of ceramic material?

SEVEN ASTRONAUTS AND ONE NAME

In November 1958, the aviation doctors of the Space Task Group presented their criteria for the selection of spaceflight candidates. They recommended that a meeting be held with representatives of industry and the various branches of the American military to select a pool of about 150 men. From this pool, thirty-six candidates would be selected for physical and psychological tests, which would ultimately lead to twelve candidates for a nine-month training and qualification period. In the end, six of these twelve men would actually fly.

There was nothing wrong with the plan, but because of the time pressure it was considered too long-winded and was discarded. Instead, during the 1958 Christmas holidays, President Eisenhower decided that the existing pool of military test pilots should offer a sufficiently broad basis. As the manned satellite project would also include classified content (such as technical data about the Atlas ICBM), for national security reasons alone the military test pilots seemed the best-qualified group of people.

A meeting of the leaders of the Space Task Group was held at NASA headquarters at the beginning of January, during which the criteria for this group of people was established. The list was short and contained precisely seven points:

1. Under forty years old
2. Maximum height of 5 feet, 9 inches
3. Excellent physical shape
4. Minimum of a bachelor's degree
5. Completion of a military test pilot academy
6. Minimum of 1,500 hours flying time
7. Qualified jet pilot

A check of the Pentagon's personnel files revealed that 110 men met these conditions. They came from three of the four branches of the military. There were five Marines, forty-seven navy pilots and fifty-eight air force pilots. A selection committee decided to form three groups and invite them to interviews. Group 1 with thirty-five men was ordered to Washington for the second of February. Twenty-four were enthusiastic about the idea and wanted to take part in Project Mercury. Six of the pilots were too tall, the rest decided against taking part. The next group, made up of thirty-four persons, came the following week. Thirty-two of them wanted to take part. Thus after inviting two groups they had fifty-six suitable candidates. This unexpectedly high consent rate made it unnecessary to invite the remaining forty-one men.

The fifty-six now had to undergo their first series of a written test, technical interview, psychiatric tests and a thorough medical examination. At the beginning of March 20, more men were eliminated. The remaining thirty-six pilots were asked to undergo additional, very rigorous medical tests at the Lovelace Clinic in Albuquerque. This prospect did not appeal to everyone, and so four more candidates dropped out.

The remaining thirty-two candidates all passed the intensive tests. The next step was mental and physical tests, which were to be carried out at the Wright Air Development Center in Dayton, Ohio. Certain that some of these tests would exceed the personal limits of some of the candidates, each of them was assured that the results would not appear in their personnel files or endanger their future military careers.

For a week, from morning till night each candidate underwent individual tests in five separate categories to test the health of the aspiring astronauts. In addition, more than thirty different laboratory tests were carried out on each candidate. The thirty-two applicants experienced what was probably the most thorough medical testing that had ever been carried out on a living human being. Unbelievably, despite the rigorous tests, the test subjects proved to be so healthy that only one did not need to report for the following tests.

For phase four of the selection program the candidates went to the Aeromedical Laboratory of the Wright Air Development Center. Awaiting them there was a sophisticated series of behavioral studies, physical endurance tests, anthropomorphic measurements and psychological tests. The individual physical and psychological limits of the individual applicants were also determined there. They had to work in pressure suits, balance on tilting tables, blow up balloons to exhaustion and run on treadmills. They were placed in isolation, sound and pressure chambers and in between had to endure more than a dozen psychological experiments and personality tests.

This last group of thirty-one was so good that even after these tests eighteen men remained. So they reflected again on the men's qualifications and professional experience and went through their service records once more. Once again, however, they failed to reach the magic number six. Even using the hardest criteria, seven men remained.

It now fell to James Donlan, the deputy administrator of NASA, to telephone each of the seven remaining candidates. He asked them if they

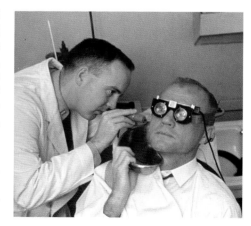

This was one of the more harmless tests during the astronaut selection process. A doctor tests John Glenn's sense of balance by pouring cold water into his ears while simultaneously checking his eye movements while wearing special glasses.

This photo of the Mercury 7 was taken four years after they were selected for the program. They are, from left to right: Gordon Cooper, Walter Schirra, Virgil Grissom, John Glenn, Deke Slayton and Scott Carpenter.

were still available for the job. All said yes, and in mid-April 1959, the new members of the American space agency were introduced to the American public. On May 1, 1959, they began their service as NASA astronauts.

From the US Marine Corps there was Lieutenant Colonel John Herschel Glenn Jr., from the navy lieutenant commanders Walter Martin Schirra Jr., Alan Bartlett Shepard Jr. and Lieutenant Malcolm Scott Carpenter, and from the U.S. Air Force captains Donald Kent "Deke" Slayton, Leroy Gordon Cooper Jr. and Virgil Ivan Grissom.

The project now needed a catchy name, especially for the public. The decision was made in late autumn. The Space Task Group had already sent proposals for possible mission emblems to NASA headquarters. One had as its theme Phaeton,

in mythology the son of the Greek sun god Helios and patron of science and research. Another had the official seal of the USA as its theme, over which were three orbital paths, and a third showed a globe, also with three orbital trajectories above it.

None of the proposals impressed the leadership in Washington. Robert Gilruth's suggestion to name the plan Project Astronaut, because it was already called that in many papers, also fell through. They found that too much emphasis was being placed on the astronauts who were to fly in the capsule. Abe Silverstein finally selected the name Mercury. The Greek messenger god Mercury was thoroughly familiar to Americans. At that time there was a popular automobile with that name, a chemical element was called that, and Mercury's Greek equivalent Hermes was known from advertisements and company names.

Moreover Mercury was the grandson of Atlas, he wore winged sandals and a helmet and thus had a symbolic import that could not be overlooked.

The Americans did not take mythology so literally anyway. That he was also the god of salesmen, traders and thieves didn't bother anybody. On December 17, 1958, the fifty-fifth anniversary of the first powered flight by the Wright Brothers, in Washington Keith Glennan announced that the USA's manned satellite program had been named Mercury. Since then, this day has been regarded as the official start of the first American manned spaceflight program, even though work on it had already been underway for more than a year.

TECHNICAL BASICS

On January 2, 1959, the Soviet Union announced the successful launch of the lunar probe Lunik 1. Together with its upper stage, the space vehicle weighed almost 3,306 pounds, and it was the first spacecraft to reach so-called second cosmic speed and leave the earth's field of gravity. Although the probe missed its actual target, it was nevertheless the first manmade object to become an artificial satellite of the sun.

One month earlier NASA had also attempted to launch a moon probe, Pioneer 3. It failed to achieve the required speed, however, and after thirty-eight hours fell back to earth. Much bitterer for the Americans, however, was the weight difference between the two vehicles. The American probe weighed less than thirteen pounds. Fifteen months after Sputnik 1, the feeling of lagging behind the Soviets was still constantly present.

In the spring of 1959, the matter of the heat shield had still not been settled. The tenders for the capsule had still been selected on the basis of the beryllium heat sink. In January 1959, however, the contract with McDonnell was modified so that the Mercury spaceship could accept both alternatives. The company was to deliver six to eight capsules with ablative shields and four to six with beryllium heat sinks.

Faith in the beryllium solution dropped, however, when it was found that there were just two manufacturers in the United States that could forge the material on an industrial scale. Only one of them, Brush Beryllium of Cleveland, was capable of making large spherical segments. Another point

This program symbol is located near the site of the former Launch Complex 14. It represents the seven original Mercury astronauts embedded in the astronomical symbol for the planet Mercury.

was that, when it returned from orbit, the Mercury capsule had to enter the upper atmosphere at a very shallow angle, in order to maintain bearable pressure loads on the pilots. Consequently, however, this resulted in a long flight through the atmosphere and thus a longer period of high temperatures. There was a danger that the white-hot beryllium would gradually transfer its heat to the pressure cabin and cook the occupants as if they were in a pressure cooker.

It thus began to emerge that it might only be possible to use the beryllium cooling body for the suborbital flights with their much lower thermal loads.

The ablation technique was also stuck in its infancy, however. The Jupiter C test of 1957 had shown that the system obviously worked during the steep and brief ascent of a military ballistic missile. Whether it was also suitable for the shallow reentry angle of a manned spacecraft with its lengthy period in the high-temperature zone remained to be seen, however.

The question could not be answered theoretically or in the laboratory. The feasibility of the ablative heat shield for manned orbital flights had to be tested, and it would have to be done with Big Joe. In NASA headquarters, Keith Glennan and Abe Silverstein decided, however, that development of the cooling body heat should continue as insurance against the possible failure of the ablative heat shield.

Another critical question was whether or not the heat shield should be jettisoned before landing. With the principle of the heat sink, no other solution would have been possible. It would have been too dangerous for the pilot to allow it to remain on the space vehicle after the thermal phase of the reentry. In the event of a landing on solid ground there would also have been the danger of setting fire to grass or forest. On the other hand, a jettisonable shield would bring added complexity to the overall system and involve the risk that it might separate before it had accomplished its primary task.

From 1958-1961, Abe Silverstein was head of NASA's manned spaceflight program.

Problems like these lasted through the first half of 1959. The more detailed the design of the Mercury capsule and the clearer its interface with the launch vehicle became, the more obvious were the gaps in knowledge that had to be filled. The Langley Center carried out five different experimental programs to close these gaps.

A series of tests involving the aerodynamic behavior of the capsule in freefall and under the drogue chute had been under way since the summer of 1958. By January 1959, more than 100 drops had been made from aircraft. The tests used steel cylinders the exact shape and size of the Mercury capsule, filled with concrete to reproduce its precise weight and center of gravity.

A second project group had the task of defining the layout of the rescue system. This group worked on Wallops Island. The first "live firing" of the rescue tower took place on March 11, 1959. It used a boilerplate, a very simple capsule model that only simulated the shape, size and weight of the Mercury capsule but was otherwise functionless. A modified Recruit rocket was used to propel the rescue tower. The test simulated a rescue from a rocket still on the ground, a so-called "launch pad abort."

The experiment was a spectacular failure. The rescue tower ignited, pulled the capsule wagging behind it, finally pitched forward, made two loops and then crashed onto the beach. The flight was such a disaster that the engineers at Langley spent several weeks considering whether the tractor system should not be replaced by a pusher system. An intensive investigation revealed that there had been a manufacturing error in one of the three nozzle openings. They also recognized that the tower-capsule's tendency to topple was significantly improved if they mounted the tower about three centimeters off the centerline.

A third series of experiments took place in the wind tunnels at Ames and Langley. They concerned the capsule's stability in the hypersonic range at speeds of Mach 22 down into the transonic range, in all atmospheric conditions and at all angles of attack. There was still empirical data on hand for the truncated cone design, especially concerning vibration levels, however, almost nothing was known about flutter behavior and changes in center of gravity, and so the wind tunnels in Sunnyvale and Hampton rumbled day and night.

A fourth program concerned how to deal with the shock of landing. Ideally the impact velocity beneath the parachute should not exceed about thirty feet per second. This presented no difficulties if the landing took place in water as planned and precisely in the vertical. The problem was that out in the open they could never expect a precisely vertical landing and had to anticipate, in some cases, significant lateral speeds. And there was always the possibility that the capsule might come down on land after an aborted launch or emergency landing, which would have led to a much harder landing impact.

The solution was a jettisonable heat shield with integrated damping cushion. The heat shield would be jettisoned from the structure just before contact with the surface. In doing so it pulled out a sort of air bag that was attached to the heat shield on one side and to the capsule on the other. The whole thing produced an accordion-like structure, a little over a yard long, which could efficiently damp high descent speeds.

A fifth problem area was the parachute system. In the spring of 1959, neither the stabilizing parachute nor the main parachute was qualified for their role in Project Mercury. It was realized that there was almost no data concerning the opening behavior of parachutes at high speeds and altitudes.

There was an ugly incident on April 1 involving a modified cargo parachute with an enlarged opening skirt, when the canopy collapsed and the boilerplate capsule crashed. It was subsequently decided to abandon this design and proceed with a ribbon parachute with a diameter of just under sixty-five feet.

Aside from the basic research, which had to be done in haste, there were many other problem areas, such as formulating the landing and recovery procedures as well as setting up a worldwide tracking and radio monitoring system.

Concerning the first point, from the beginning the Mercury planners assumed that they would involve the US Navy. The navy was not averse to taking on the job and it could be assumed that it would be capable of doing so on account of its vast experience in search and rescue operations. The problem, therefore, was not the navy, rather the many precautions that had to be built into the capsule to guarantee a safe water landing and recovery. It must not tip over in rough seas, and it had to be equipped with special radio and signaling devices and with fittings to enable it to be picked up. It also had to be capable of coming down on land in the event of an emergency.

As to the communications and radio-tracking network, it was hoped that they would be able to use the infrastructure created for the International Geophysical Year. This hope was deceptive, how-

The recovery of the capsule Freedom 7. The capsule's pneumatic landing bag can easily be seen.

ever. The so-called Minitrack Network, which had been built north and south of the 75th parallel in the Western Hemisphere for Project Vanguard proved almost worthless for Mercury. Military communications installations were more widely spaced and equipped with much lower bandwidths than hoped.

March 17-18, 1959, were decisive dates for Project Mercury. On those two days, in its St. Louis factory McDonnell presented the first detailed full-scale mockup of the manned Mercury capsule to the Space Task Group. This event was the first significant contract milestone and was titled Mockup Review Inspection.

The mockup could be broken down into seven main components: the adapter ring to the launch vehicle, the retrorocket pack, the heat shield, the pressure cabin, the outer airframe,

McDonnell even displayed its Mercury mockup at an open house in 1959.

the antenna canister and the pylon with the rescue rockets.

Designers and managers surrounded the model. They came from the Space Task Group, from NASA headquarters, from the government and from the air force and navy. All in all there were more than thirty guests present, including a Marine test pilot by the name of John Glenn, who had been sent by the Navy Bureau of Aeronautics. There were also about forty people from McDonnell.

The meeting led to forty proposals for modifications to the capsule, of which twenty-five were immediately accepted. The rest were given to working groups for further study. The most important changes were the addition of a side exit hatch, blinds for the windows, reinforced surfaces on which the pilot could climb up and out through the cylindrical nose cone and a camera, with which the pilot was to take photographs throughout the flight.

HIDEOUT ON THE CAPE

From today's perspective it is hard to imagine, but until 1959 there were virtually no NASA facilities in Cape Canaveral. The new space agency had no laboratories there, no integration sites, no halls and hangars, not even an office and certainly no launch installations. The area was the scene of hectic activity. The army, navy and air force maintained extensive installations of all kinds. "Missile row," the long chain of launch pads on the beach up to Merritt Island, was in constant upheaval and construction work went on day and night.

The Space Task Group's very first employee on the Cape was Porter Brown. On May 1, 1959, he moved into an office in Port Canaveral. His first job was to understand the confusing relations in this beehive and make himself known there.

By mid-1959, there were already more than two-dozen launch pads on the Cape. More were constantly being built, and yet there were never enough. There were literally just as many military and civilian organizations there as launch pads and all jealously guarded their domains. As well, there were rigorous military secrecy rules. No one there had any use for a new player and no one had even the slightest desire to relinquish space or resources to Project Mercury. Even the only previously existing NASA office on the Cape, the Jet Propulsion Laboratory, was one of the established top dogs. Together with Wernher von Braun's team, the staff of the JPL were already in the process of finding additional space in order to modify the Jupiter C launch facilities for the new Juno II so that the next series of Explorer satellites could be launched.

The Space Task Group had actually been assigned Hangar S in the Cape's industrial area. There was just one catch: the Naval Research Laboratory Team, which had built the structure for

Hangar S as it looked in the early 1960s.

33

Launch Complex 14 at Cape Canaveral in May 1963. On the launch pad is Atlas number 130 carrying Faith 7.

the Vanguard program, was still there, and the project group had no intention of leaving, because there were two launches left and they were still hoping for a follow-up project.

In June Porter Brown finally succeeded in acquiring a corner in Hangar S marked out with barrier tape, but not before he had been told emphatically by the "masters of the house" not to step over the barrier tape under any circumstances. That was just in time, because that same month thirty-five engineers and technicians arrived that desperately needed the space to prepare the capsule for the Big Joe flight attempt.

Porter Brown already had the next problem around his neck: the air force had assigned NASA Launch Complex 14 for the Big Joe launch. At the time it was one of a total of five launch pads for the Atlas. At exactly the same time that the Big Joe launch was to take place, however, it was occupied for a MIDAS mission by the air force.

Although it was clear to everyone that the launch plans for the Mercury qualification program had a "very provisional character," to put it mildly, the Space Task Group wanted assurance that Launch Pad 14 would be reserved exclusively for the Mercury Atlas launches and would not be hogged by other programs. This suggestion was rejected out of hand by the commander of the Air Force Missile Test Center, Major General Donald Yates. He explained that in case of maximum utilization of the facilities on the Cape, he would allocate the next available pad to a scheduled Mercury Atlas launch. The conflict made its way up the command structure, and the Secretary of Defense ultimately came down on the side of General Yates. As it turned out in the end, however, the fuss was for nothing, for as the Mercury Program gained strength, NASA's influence on the Cape became ever stronger. In the end, all of the Atlas-Mercury launches were carried out from Launch Complex 14.

SPAM OR IMPORTANT COMPONENT?

Concerns about the reliability of ballistic rockets did not originally rise from concerns about the safety of a man. The main criteria were maximum range and accuracy. The warhead had no sense anywhere else but the target. If the launch vehicle failed, the warhead could also be destroyed. Saving it made no sense. Redundancy was achieved by launching as many as four rockets at the most important targets, and at least two at less important ones.

On the other hand a man had to be brought to safety if the launch vehicle failed. The rockets had to be "man rated." The Redstone team had different ideas than Convair's Atlas people as to how that was to be achieved. The rocket people from Huntsville were handicapped by their work on Project Adam.

They regarded the astronauts as completely passive payload and insisted that no possible intervention by the passenger be allowed.

In fact there had been several interesting solutions proposed in Project Adam. If a launch was aborted on the pad, the capsule would be catapulted sideways out of the rocket and hurled into a water tank next to the launch pad. The error recognition system was based on the consideration that, because of the enormous number of critical components in the rockets, the entire launch vehicle could not be plastered with sensors. This would neither increase nor lower system safety, as thousands of sensors would have been further potential sources of error.

The conclusion was reached, however, that many failure modes would lead to the same results. It was therefore necessary to install sensors for observing the results (for example loss of pressure in the engines or voltage changes in the electrical system). Thus they were able to limit the kinds of sensor to a few types and their number to several dozen. Because of this, however, the cause of the error, a crack in the combustion chamber or a leak in the fuel tank, was causally separated from the measurement of the anomaly and therefore no longer assessable by the pilots. Moreover, the Redstone people argued that most rocket failure modes happened so quickly that any action by the astronaut would have been impossible. For this reason, they concluded, it would be pointless to integrate him at all.

This fundamental consideration touched on the general role of the space traveler in the capsule, which in 1959 was still unresolved. It was the view of the astronauts that the presence of a human improved the reliability of the system. When asked why he had not applied to become an astronaut, Charles Yeager, the first man to break the sound barrier, answered: "I did not want to become spam in a can." This meant that he did not want to be a passive, functionless element in this system. Others, including John Pierce, head of development at Bell Telephone Laboratories and one of the fathers of the communications satellite, had similar thoughts. In 1960, he observed rather spitefully: "The last thing we need in Project Mercury is a highly-trained space pilot constantly tampering with the controls."

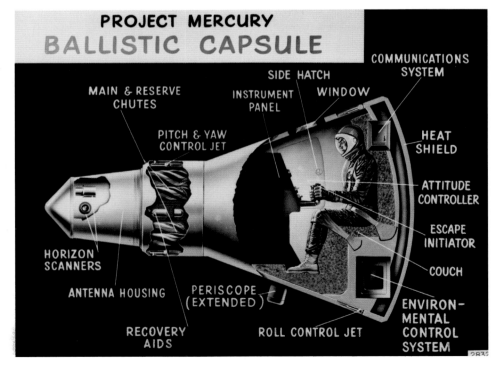

This contemporary illustration from 1959 stresses very vividly the "active" astronaut.

This was not a topic in the Space Task Group, however. They were convinced that the astronauts should be given much more than the role of passenger. There the pilots were seen as an important element that could determine the success or failure of a mission. In fact, however, the role of the pilots gradually became clearer after the astronauts became actively involved. In the end the pilot was an indispensible element in the function chain.

At Convair, maker of the Atlas, they approached the matter much more pragmatically than the Redstone people. There, the Abort Sensing and Implementation System, as it was called, was simply regarded as a kind of "accessory," special equipment for the Atlas rockets purchased by NASA, if they wanted it. This way of thinking, separate from the pilots, eased the work considerably.

One of the legacies of Project Mercury was the change in this philosophy during the course of the program. In the beginning, man's capabilities in space were unknown. For this reason, the system had to be created so that it could function without him. In the course of the program this point of view changed completely, for on several missions the success of the mission was assured by men who could fly the spaceship manually.

A QUESTION OF RELIABILITY

The Atlas rocket had approximately 40,000 so-called "critical parts," and there were 40,000 more in the systems of the Mercury capsule. One of the standard methods of creating reliability is to introduce redundancy. This involves supplementing a component that—for whatever reason—is not trusted, with a second component with the same function, but preferably of a different design. Another method is to simplify systems, subsystems and components as much as possible in order to reduce complexity and thus their susceptibility to failure.

Joachim Küttner, a former German test pilot who had joined Wernher von Braun's team in the early 1950s, was responsible for "man rating" the Redstone. His first model was the Jupiter C, a variant of the Redstone, which had been developed for tests with the reentry system of the Jupiter medium-range missile. Küttner began with a simplification: he eliminated the missile's stage separation equipment and a number of components that were only required by the variants of the Jupiter C used in experiments. He did, however, retain the longer fuel tank of the Jupiter C. This gave the Mercury-Redstone twenty seconds more burn time. This was necessary, because a non-toxic mixture of 75% ethyl alcohol and liquid oxygen was to be used to fuel the Mercury-Redstone instead of the ten-percent more energy-rich but poisonous Hydyne, the fuel used for the Jupiter C.

Then he had to add complexity in order to be able to place the Mercury capsule atop the rocket. In particular, the rocket needed an adapter ring to enable a clean separation between rocket and capsule. It also needed an added nitrogen pressure tank with which to pressurize the larger fuel tank. And it needed a second hydrogen-peroxide tank to provide enough fuel for the turbopump, for its running time had been extended.

Joachim Küttner (far left) was responsible for "man rating" the Redstone. This photo shows the rocket that carried Virgil Grissom into space on mission MR-4. The astronaut himself is third from the right.

Then Küttner eliminated complexity again. He replaced the very modern but complicated ST-80 autopilot of the Jupiter C and reinstalled the simpler LEV-3 inertial navigation system of the original Redstone. He was also able to eliminate the Terminal Guidance System, which controlled the Jupiter C's precisely defined ascent angle. Then the previous model of the A-7 engine had to be replaced with the newest variant of this engine to prevent a possible spare parts problem later in the program. And so it went on and on.

By July 1959, the number of modifications was approaching 800. The danger that they were opening Pandora's Box, with one change leading to the next, was great. Already the reliability level of the Jupiter C was not comparable to that of the Mercury-

Redstone, because it had become a variant unto itself. They could only hope that the combination of all the measures taken produced a product that was at least as reliable as the original model.

Unlike the rockets, the engineers could not fall back on an earlier base model. The only possible recipe for maximum possible reliability was therefore redundancy. Altogether more than sixty different redundancies were designed into the capsule. They were supposed to enable the pilot to successfully complete a flight despite the failure of the primary system—or simply survive.

In August 1959, McDonnell estimated the overall probability of achieving the mission objective in a suborbital Mercury-Redstone mission at 77.8

percent. They estimated the probability of the pilot surviving the mission at 99.1 percent.

It was a different matter with the Atlas. A few test missions had been carried out in August 1959 and roughly every second one ended in failure. In November 1959, the Air Force Ballistic Mission Division informed the Space Task Group that it anticipated a reliability figure of about seventy-five percent by roughly mid-1961. By mid-1962, it declared, a figure of about eighty-six percent could be achieved. Understandably, NASA was little impressed by these figures.

The result of this was that the Space Task Group formed its own working group to examine the operating reliability of the Redstone and the Atlas. Its members were pleasantly surprised, however, when they found that among all the failure modes there were relatively few catastrophic scenarios that developed extremely quickly. Most scenarios developed over the course of at least several seconds, enough time to activate the capsule's escape system.

The working group identified the greatest problem as the danger of so-called "nuisance aborts," various inadvertent flight aborts caused by the overemphasis on pilot safety, by too many or overly sensitive sensors or the installation of too many redundancies, which added complexity to the system. The working group decided that all signals coming from the abort sensors should be displayed on the pilot's instrument panel. Parameters that were only critical but not catastrophic must in no case trigger an automatic abort. Situations considered "critical" included a partial loss of thrust, fire in the capsule, deviations from the predetermined flight path, a drop in fuel tank pressure, and other similar occurrences. Basically they were all situations that were not immediately life threatening for the astronauts or could possibly be corrected during a later phase of flight.

"Catastrophic" failures, on the other hand, were defined as leaving no time for corrective actions or a manual flight abort. Typical situations of this kind included excessive deviations from the normal flight path, high roll rates, sudden loss of pressure in the fuel tanks and engines, complete loss of electrical power, total loss of thrust immediately after liftoff, and so on. If one of these situations occurred, the flight abort system had to intervene immediately.

THE FLIGHT OF "BIG JOE"

Concerns about the reliability of the Atlas were not unfounded. On April 14, 1959, the Air Force launched the first D-model. It exploded thirty seconds after leaving the launch pad. Of the four Atlas Ds that followed, two flights were complete failures, while the other two could only be seen as "partial successes" with much optimism. But then, in July and August, serial numbers 9 and 11, both D models, made perfect launches and the last two C models were also flown successfully.

September 9, 1959, was a decisive date, both for the Atlas program and for the Mercury project. On that day, separated by barely twelve hours, both the first significant flight-test of the Mercury program and the final qualification of the Atlas as an intercontinental ballistic missile took place. While at Cape Canaveral the attention of NASA and the Space Task Group was focused on the Atlas booster with the serial number 10D, which was supposed to begin its journey southeastwards

Little Joe 1 Boilerplate

Early Boilerplate Capsule

The illustration on the left depicts an early Mercury boilerplate. The external shapes of all the boilerplates differed from one another to a certain degree, and each was unique. The illustration on the right shows the external appearance of the Little Joe 1 Boilerplate.

1. Drogue chute container. 2. Location aid for capsule recovery. 3. Braking parachute actuator. 4. Aft capsule section. 5. Pressure cell. 6. Cable attachment. 7. Marman clamp for attachment to rocket adapter. 8. Heat shield dummy or heat shield. 9. Nitrogen tanks. 10. Instruments (or bio-container). 11. Attitude control jets. 12. Pressure cabin. 13. Access hatch. 14. VHF antenna. 15. S- and C-Band antennas. 16. Main parachute container. 17. Recovery hooks (total of 3). 18. Antenna cone.

over the Atlantic Missile Range, at the Vandenberg missile base in California Atlas number 12D rose from Launch Pad 14 and headed southwest over the Pacific Missile Range.

The Big Joe flight, also called the Atlas ablation test, had actually been scheduled for the 4th of July, but then delays had arisen with the instrumentation and telemetry systems for the booster and capsule. The launch vehicle was the sixth Atlas D in total. The Mercury capsule atop the rocket was one of the "boilerplates," an almost empty shell with

measuring devices, but equivalent to a "real" Mercury in terms of weight and center of gravity. But above all, and this was the key, it was equipped with an ablative heat shield. A whole series of such dummy capsules was used in the program. Most were relatively crudely forged shells, which resembled the high-technology end product only in their shape.

The Big Joe boilerplate had been built by technicians at the NASA centers in Langley and Lewis using the same Inconel alloy that would be used for the actual capsules, for it was intended to survive a reentry. It was, however, largely empty and not pressurized, and there was no life-support system. Only for the instruments was there a small pressure vessel in the interior. The escape tower was absent as were the retrorockets.

More than 100 sensors were mounted inside and outside the hull of the capsule to measure temperature distribution. Only a small part of the data went to the ground via telemetry link. The majority was to be saved by recording devices on board the boilerplate. The capsule also had a simple attitude control system that operated with nitrogen. It was controlled by an attitude control system made by Honeywell. The system was considered mission critical, for it had to execute the 180-degree turning maneuver after separation from the carrier.

The NASA people had relatively little to do with the rockets themselves. There was only a liaison man between the NASA team and those of the air force and Convair. The job fell to the indestructible Porter Brown, who was given the title Atlas-Mercury Test Coordinator. His main job was to see to it that his people did not stick their noses into the highly secret details of the brand-new Atlas intercontinental missile.

The capsule test group itself was led by Scott Simpkinson. His team consisted of forty-five people who together with their capsule had been camped out in the segregated corner of Hangar S since the second week of June. As before, the Vanguard unit from NASA's Goddard Center occupied most of the building. They were preparing for the launch of Vanguard III, which was also supposed to take place in September.

Soon after launch, Atlas 10D was supposed to follow a shallow parabolic trajectory whose zenith was at an altitude of a good one-hundred miles. While still in the acceleration phase it was supposed to go into the descending arm of the parabola and then, at a speed of about 15,532 miles per hour, release the Mercury boilerplate. After this the capsule was supposed to fire its attitude control engine, turn 180 degrees and carry out the reentry into the earth's atmosphere.

The kinetic energy of the capsule and the drag would then shroud the capsule in a white-hot plasma envelope while it descended at very high speed into the lower atmosphere. It was hoped that the greater part of this heat would be carried away by the shock wave and the melting material of the heat shield, as planned, in order to prevent the capsule from being vaporized.

The countdown on the night of September 8-9 proceeded calmly and without problems. At 03:19, the two vernier engines ignited first, followed by the three main engines, and the Atlas lifted off.

It was a beautiful launch. The flame jets of the Atlas lit the night sky and Brevard County shook from the roar of rocket engines. Everyone was excited—at least for the first two minutes. Then the flight path plotters' suddenly flattening curve indicated that the two outer engines and the thrust frame had not been jettisoned.

Obviously all systems in the capsule were functioning normally, but the engineers were unable to record the envisaged 180-degree turn. Everything indicated that the attitude-control engine was working normally. The added weight of the booster engines still on the Atlas reduced speed to just over 621 miles per hour, and the Burroughs tracking computer on the Cape calculated a new impact point: it was 497 miles shorter than planned.

Then Atlas and Mercury disappeared into the radio blackout and a nerve-racking period of waiting began. The fate of the Big Joe-Mercury was uncertain and they would only learn more if they succeeded in recovering the capsule and reading the tape recording on board the capsule.

Six vessels of the 4th Destroyer Flotilla, which was now far from the envisaged splashdown point, headed northwest toward the test range at maximum speed. Patrol aircraft began flying search patterns. The first sign of the capsule was spotted before sunrise, when ships and ranging stations on both sides of the Atlantic registered the explosion of a Sofar bomb (Sofar = Sound Fixing And Ranging). The Mercury capsules were equipped with two of these explosive devices, each of which weighed

Big Joe, a few hours before launch.

one pound. The first was dropped when the main parachute deployed. The second remained on board the Mercury and was only supposed to explode if the capsule sank. Both Sofar bombs were set to explode at a depth of 3,937 feet.

As planned, the Sofar signal made a triangulation possible, the coordinates of which were immediately passed on to the search aircraft.

Soon afterwards a Navy P2V Neptune patrol aircraft also obtained contact with the capsule's radio beacon and soon after sunrise the crew saw the Mercury boilerplate bobbing up and down in the water. The pilots guided the nearest recovery vessel, which was still more than one-hundred miles away, to the target area. Seven hours later the destroyer Strong heaved the precious cargo on deck. The capsule's subsequent return to the Cape by water, land and air took less than twelve hours. As soon as the cargo plane landed at Patrick Air Force Base, the capsule was loaded onto a transport truck and, accompanied by a police escort, it began the last part of the journey: the fifteen miles through Cocoa Beach to Cape Canaveral.

By about midnight the capsule was back in Hangar S. Every single NASA member was present to receive it. Robert Gilruth and Max Faget from Space Task Group management had come. Then

someone removed the canvas from the capsule—the heat shield was still a military secret—and all reverently inspected the heat shield. The entire group was astonished by the excellent condition the material was in. Only a thin layer had melted away. Extending from the center of the shield, a star-shaped pattern of glass beads extended over the shield. The rear part of the capsule was a little singed, and the words United States, which were painted white, had become only slightly discolored.

The exhausted but happy crew unscrewed the two halves of the pressure vessel inside the capsule and passed to Gilruth a letter they had placed there before launch in anticipation and hope of just this moment. The text read: "This note comes to you after being transported into space during the successful flight of the "Big Joe" capsule, the first full-scale flight operation associated with Project Mercury. The people who have worked on this project hereby send you greetings and congratulations."

A few days later the story of the Big Joe mission was reconstructed from the saved onboard data. As suspected, the booster rockets had not separated from the launch vehicle and after burnout were pulled along by the sustainer as dead weight. As a result, the terminal velocity of the Atlas was 621 miles per hour too low, which resulted in the flight path becoming a steeper parabola with a lower vertex than planned. This was part of the reason why the capsule was initially unable to separate from the rocket after sustainer burnout. It was simply pressed further into its bedding by the incoming atmosphere as it began descending into the upper atmosphere. In its futile attempt to turn itself and the still attached rocket through 180 degrees for reentry, the capsule had used up all its fuel in a very short time. At an altitude of just over sixty-two miles, the effects of the atmosphere were so great that it finally broke loose, two minutes and eighteen seconds after the time actually envisaged and at a speed of over 14,289 miles per hour.

Thanks to its carefully sited center of gravity, however, the capsule turned itself around in the

Mission Data	
Mission Name	Big Joe 1
Date	September 9, 1959
Launch Site	Cape Canaveral, Launch Pad 14
Launch Vehicle	Atlas D (Number 10D)
Spacecraft	Boilerplate
Spacecraft Weight	2,558 lbs
Flight Path	Suborbital
Maximum velocity	14,856 mph
Flight duration	13 minutes
Flight distance	1,429 miles
Flight path apex	95 miles
Landing site	Central Atlantic
Recovery ship	USS Strong

growing air stream even without the help of thrusters and entered the lower layer of the atmosphere in the proper position. The duration of the maximum heat load was shorter than originally planned, but because of the steeper descent the temperature was higher. The sequence of events during reentry and splashdown, the structure, the instrumentation and the cooling system had worked well. The results of this mission fostered such confidence that the backup mission by Big Joe II was cancelled within three weeks.

When the final report on the Big Joe test with lessons learned was presented at the end of October, it immediately led to a number of design changes in the capsule and to changes in program planning.

After this mission, the beryllium heat sink was no longer pursued for orbital flights. Particularly satisfying was the fact that the capsule had achieved the correct descent attitude as a result of the position of its center of gravity, without any active attitude control. In a nutshell: the members of the Space Task Group were very enthusiastic.

Less enthusiastic, however, were the Atlas people. Booster 10D had failed during stage separation and therefore the mission was rated a complete failure by the air force. Fortunately, on the other side of the country at Vandenberg, Atlas 12D had functioned perfectly in its flight of over 5,591 miles just hours later and had passed operational qualification.

LAUNCH PROBLEMS FOR LITTLE JOE

The mood in Project Mercury had been considerably worse two and a half weeks before the launch of Big Joe. An extremely baffling anomaly had occurred during the very first planned use of a launch vehicle, and it was only through very good fortune that no one had been injured.

On the morning of August 21, 1959, a Friday, the first Little Joe, designated LJ 1, sat on its launch pylon, which was angled east towards the sea. At its tip was a Mercury boilerplate without parachute system but with an active escape tower.

The objective of this first mission was to test the escape tower under a load calculated as the maximum for a Mercury-Atlas launch. Evacuation of the launch site began thirty-five minutes prior to ignition. The batteries for the flight sequence programmer and the self-destruct system were just being loaded when suddenly there was a large flash, followed by a mighty roar.

Photographers and technicians scrambled for cover, but then it was all over. When the bewildered observers raised their heads, they saw that the escape tower and capsule had gone off on their own. They had simply launched. The Little Joe was still intact on the launch pylon. A look up showed what was happening, for by then the capsule had reached the apogee of its parabolic path at a height of about 2,132 feet. At that instant the escape tower's clamping ring released, a small solid-fuel rocket separated it from the boilerplate and, twenty seconds after the unintentional launch, the capsule crashed into the water off the beach.

The accident report appeared on September 18. The accident was blamed on a coil that had been retrofitted in the capsule. It had been added as a positive redundancy, in order to better protect the life of the test animals that were to be used on coming flights. This circuit was supposed to

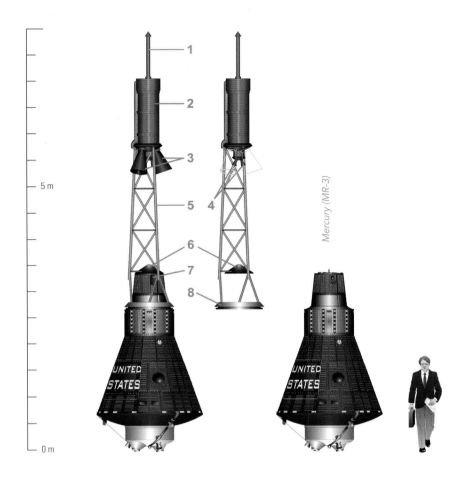

5 m

Mercury (MR-3)

0 m

UNITED STATES

UNITED STATES

The escape tower was a rescue device for the astronauts in case of a catastrophic failure of the launch vehicle.

1. Aerodynamic tip. 2. Rescue rockets. 3. Escape rocket nozzles. 4. Separation rockets for the escape tower (hidden behind the nozzles). 5. Escape tower framework structure. 6. Antenna cone cover. 7. Horizon seeker cover. 8. Escape tower separation ring with aerodynamic fairing.

prevent exactly what had happened, namely the unintentional firing of the escape tower.

On September 25, 1959, North American transported the sixth and last Little Joe airframe to Wallops Island. At the beginning of October, therefore, the Space Task Group had available all the Little Joes it had ordered. Later that autumn three of the rockets were launched at precise intervals of one month.

The series began on October 4 with the same launch vehicle that in August had so ignominiously ditched its Mercury capsule. After the experiences of the summer, the Space Task Group had become cautious and scheduled this flight, designated LJ-6, as a pure test mission. Consequently this first Little

Mission Data	
Mission Name	Little Joe 1
Date	August 21, 1959
Launch Site	Wallops Island
Launch Vehicle	Little Joe
Spacecraft	Boilerplate
Spacecraft Weight	2,558 lbs
Flight Path	Suborbital
Maximum velocity	186 mph
Flight duration	20 seconds
Flight distance	2,625 feet
Flight path apex	2,000 feet
Landing site	Beach, Wallops Island
Recovery ship	USS Strong

Mission Data	
Mission Name	Little Joe 6
Date	October 4, 1959
Launch Site	Wallops Island
Launch Vehicle	Little Joe
Spacecraft	Boilerplate
Spacecraft Weight	2,502 lbs
Flight Path	Suborbital
Maximum velocity	3,706 mph
Flight duration	4 minutes, 10 seconds
Flight distance	79 miles
Flight path apex	37 miles
Landing site	western Atlantic
Recovery ship	USS Strong

Joe carried just a double dummy: a boilerplate Mercury without instrumentation with an inactive escape tower. The modest goal of the flight was to test the reliability of the Little Joe. The eight solid-fuel rockets were to be ignited sequentially. This would be followed by the ignition on the ground of two Pollux engines and the four Recruit boosters. Just before burnout, about thirty seconds after the vehicle left the launch pylon, the first two sustainers would fire, after which the two remaining Pollux boosters would be activated.

Little Joe 6 ignited and climbed to a height of thirty-seven miles. Then it was intentionally destroyed to test the self-destruct system. The wreckage fell a good seventy-nine miles off the shore of Wallops Island. The experiment had been a success.

Flight number two was supposed to carry out the mission originally planned for August 21 and therefore was given the designation LJ-1a. In addition to testing the escape tower, the parachute deployment sequence and subsequent recovery were also to be tested. The escape tower's Grand Central rocket motor was to be fired at the moment of maximum dynamic load. Abbreviated max q, this area is the flight zone in which the launch vehicle and spacecraft encounter the maximum amount of drag. The Atlas and the Little Joe reached this point more or less at the same point, sixty seconds after liftoff. Under these conditions the combination was exposed to a pressure of about five tons per square yard.

A surprisingly large number of news people came to Wallops Island on November 4, 1959, to watch the launch. The launch took place as scheduled and appeared to go off completely smoothly. The rocket obviously followed the planned flight path, the rocket separated the escape tower from the capsule, the stabilizer and main parachutes deployed as anticipated, and splashdown took place eleven miles off the coast.

It had not escaped the test engineers, however, that, for reasons unknown, the escape tower had not fired until ten seconds after it had passed the point of maximum pressure. Why this happened was never discovered. An analysis only showed that, because of the delayed ignition of the escape tower, only a tenth of the anticipated dynamic pressure was achieved. It was clear that the test would have to be repeated, but it could not be done on the next Little Joe flight, for preparations

Little Joe 1 after the capsule with escape tower separated on its own.

were well advanced and it had a completely different task.

In May 1959, the Space Task Group, together with the School of Aviation Medicine, had begun planning a Little Joe flight dedicated to space medicine purposes. A bio-pack was prepared for the mission containing spores, tissue cultures, fertilized eggs, cereals, and, most prominently, a rhesus monkey named Sam. Sam had got his name simply from the first letters of his alma mater, the School of Aviation Medicine. All of these biological specimens were housed in a pressure vessel inside the boilerplate capsule. The scientists hoped, in particular, to learn about the effects of weightlessness on Sam.

Shortly before noon on December 4, 1959, the four Recruit booster rockets and the first two Castors fired. The second set of Castors burnt out at an altitude of nineteen miles, the escape tower ignited as planned, the capsule separated from the rocket, giving it an additional impulse on its journey. The tower then separated from the capsule and the two components followed a parabolic path that reached a height of fifty-four miles at its apex. The maximum height was something of a disappointment, as at least nineteen miles more had been expected. They

had, however, failed to sufficiently take into consideration the turbulence loss caused by Little Joe's big fins. Nevertheless, Sam survived more than three minutes of weightlessness.

He survived the relatively mild reentry, the rough landing on the water and the six hours he had to spend in his pressure chamber in reasonable condition. He was fished out of the water by a destroyer and released from his cabin.

The flight was a complete success, even though it had failed to reach the desired altitude. It had been demonstrated that the rescue system worked under moderate conditions. What was still lacking was a test of the escape system under the toughest possible conditions, with maximum dynamic pressure on the rocket and capsule. Proof of this was to be obtained on the third try, mission Little Joe 1b.

NASA tried to play down the importance of this flight a little, but this was not so easy, for this mission also had a star: Miss Sam, the female counterpart to Sam. Her life was at stake and that was reason enough for the press to report at length on the short flight, which in turn caused some qualms at NASA.

To the relief of the space officials, the mission was perfect. Powered by two Pollux rocket engines and four Recruit booster rockets, the fourth Little Joe hissed into the sky on January 21, 1960. The escape system was activated at a height of nine miles and a speed of 2,050 miles per hour. Enclosed in her tiny pressure cabin, Miss Sam not only survived the flight but also performed the tasks given her. She had to watch a light, and whenever it lit up move a lever. The capsule reached an altitude of ten miles and splashed down twelve miles off the coast of Wallops Island. Within a few minutes it was sighted by a helicopter and picked up. Forty-five minutes after launch, Miss Sam was back on Wallops Island, where she proved to be in excellent condition.

At that point in the test program the Space Task Group had undertaken five launch attempts, expending five Mercury boilerplates and four of the

Mission Data	
Mission Name	Little Joe 1a
Date	November 4, 1959
Launch Site	Wallops Island
Launch Vehicle	Little Joe
Spacecraft	Boilerplate
Spacecraft Weight	2,227 lbs
Flight Path	Suborbital
Maximum velocity	2,019 mph
Flight duration	8 minutes, 11 seconds
Flight distance	12 miles
Flight path apex	9 miles
Landing site	western Atlantic
Recovery ship	USS Strong

Mission Data	
Mission Name	Little Joe 2
Date	December 4, 1959
Launch Site	Wallops Island
Launch Vehicle	Little Joe
Spacecraft	Boilerplate
Crew	rhesus monkey Sam
Spacecraft Weight	12,227 lbs
Flight Path	Suborbital
Maximum velocity	4,473 mph
Flight duration	11 minutes, 6 seconds
Flight distance	194 miles
Flight path apex	54 miles
Landing site	western Atlantic
Recovery ship	USS Borie

Mission Data	
Mission Name	Little Joe 1b
Date	January 21, 1960
Launch Site	Wallops Island
Launch Vehicle	Little Joe
Spacecraft	Boilerplate
Crew	rhesus monkey Miss Sam
Spacecraft Weight	2,227 lbs
Flight Path	Suborbital
Maximum velocity	2,057 mph
Flight duration	8 minutes, 35 seconds
Flight distance	12 miles
Flight path apex	9 miles
Landing site	western Atlantic
Recovered by	Marine helicopter

six boosters delivered by North American. Just two units were left with which to complete flight qualification tests. North American had, however, produced seven units and not six.

One of these had been used for static load tests. The Space Task Group now asked North American to also bring this airframe to flying condition and deliver it to NASA, increasing the number of available boosters to three. The success of Little Joe 1B in January 1960 meant that the sixth flight would be the first in which a "proper" Mercury from McDonnell production would fly and not just a boilerplate made in the NASA workshops.

Miss Sam takes off on her brief and physically strenuous mission.

THE SPACE SUIT

Until the end of 1959, while Project Mercury was behind schedule, it was moving forwards on a broad front. Hopes were not all that high with respect to the race with the Soviets. As before, the east continued to sail forwards in the vastness of space. On October 7, the Soviet space probe Lunik III had become the first space vehicle to photograph the far side of the moon and once again America was left empty handed.

At least the American astronauts were now in full training. The press had already acclaimed them heroes setting out to challenge the Soviet empire in space. Their mission was already relatively well defined. They were to maintain contact with the ground stations and send regular reports, carry out scientific experiments and observations, control the attitude of the capsule in space, monitor the onboard instruments, control and fire the retrorockets, initiate emergency procedures, if necessary deploy the escape system and activate the landing parachute. The man was considered an active system component, but for this he needed a precisely defined working environment.

Creation of the "environment cocoon," which would assure the men on board the basic metabolic needs, was one of the most complex and most difficult tasks of the Mercury engineers. The safety regulations required that the life-support system should be designed on a redundant basis. The capsule was a pressure cell, in which under normal conditions there was a pressure of 300 millibars in an atmosphere of pure oxygen.

The astronaut was also to wear a space suit in case the pressure cell should be damaged for whatever reason. The spacesuit thus had the function of a second, internal pressure cell in case the first, outer cell could not complete its function.

The first approach to the space suit was to simply modify a standard pressure suit used in military aviation. This idea was quickly dropped, however. The military suits were all "high-pressure suits," designed to help the pilot endure short-term high G loads, especially in turning flight. They offered insufficient protection for high altitudes, had enormous leakage rates, inadequate air circulation and were completely unsuitable for open space. Wearing comfort was miserable, however, this was an extremely important point, for in an emergency an astronaut might be forced to spend several days in his suit. The astronauts therefore took special interest in the development of the space suit.

Three companies had applied for the contract for the space suit in June 1959: the David Clark Company from Worcester, Massachusetts, main provider of pressure suits to the Air Force; the International Latex Company of Dover, Delaware, which specialized in rubberized materials; and finally the B.F. Goodrich Company of Akron, Ohio, which made almost all the pressure suits for the U.S. Navy. On July 22, 1959, Goodrich became the first provider of space suits for the young NASA.

Astronaut Walter "Wally" Schirra was in charge of space suit development.

SEVEN MEN FOR SPACE

At the end of April 1959, the seven future astronauts began their service with the Space Task Group. They received the very first "teaching module" on April 29, 1959, when they were briefed on the working of the Mercury emergency rescue system. For the next two weeks they received instruction in every aspect of the program. The third week saw the start of a tour of the United States in which they visited every main contractor of the Mercury program and familiarized themselves with mockups, flight hardware and production processes.

Then it was the turn of the installations at Cape Canaveral. In military and medical installations

they learned to recognize their physical reactions to extraordinary loads and unusual physical symptoms. They wore pressure suits for hours, breathed high concentrations of carbon monoxide, spent hours in heat and pressure chambers and made parabolic flights in the C-131 and in the back seat of an F-100, in order to feel weightlessness for several seconds. All of the astronauts learned to scuba dive and spent many hours under water. Part of the training time was freely disposable. It was expected that they would spend several hours a week in a jet that had been provided, so that they could maintain their skills as pilots, deal with the special field they had decided on, and maintain their physical fitness.

In the late summer of 1959, each astronaut spent two weeks in the centrifuge in Johnsonville, in order to get to know the load profile during launch and splashdown. Then they went to the Cape for the Big Joe launch, to McDonnell for integration tests with the capsule, and to Goodrich to test the space suit. Meanwhile there were all other possible training modules, for example communication with the tracking network or survival training, which was followed with great interest by the press.

This first training segment still had many elements of a university seminar. In the first year of their training they received lectures in space physics, flight guidance and flight control, space navigation and space medicine. Each of the astronauts also spent eight hours in the Morehead Planetarium at the University of North Carolina, in order to learn to navigate the heavens. Each of the men selected a special program-related field, into which he could bring his professional knowledge.

Scott Carpenter was responsible for communications and navigation, as he had received special training in this field during his time in the navy. Virgil Grissom, who had a degree in mechanical engineering from Purdue University, was the expert in electromechanical, automatic and manual flight systems. John Glenn had the most flying experience and had flown the most aircraft types. He therefore became involved in designing the cockpit layout. Walter Schirra, a graduate of the naval academy, dealt with life support systems and the space suit. Alan Shepard, also a graduate of the naval academy, specialized in space tracking and recovery activities. Gordon Cooper and Deke Slayton looked after the interface with the Redstone and Atlas rockets. Before being chosen to be an astronaut, Cooper had worked as a test engineer and Slayton, who had a degree as an aeronautical engineer from the University of Minnesota, had worked as a designer at Boeing for two years before going to Edwards as a test pilot.

Assigning these tasks to the astronauts proved to be a wonderful idea. In this way they were able to have a direct influence on development of the Mercury systems, and they were constantly informed about every aspect, every problem, every advance and every failure in every system and subsystem in the entire program. The constant exchange about the status of the information received and coordination with each other had the effect of welding the astronauts into a homogenous and influential group within the program.

The second phase of the astronauts' training began with the turn of the year 1960. Its lecture character now gave way to mission training, but the astronauts continued to travel about the country, keeping track of the development and production of their launch vehicles and spacecraft. Looking back, the program manager thought these visits as perhaps the outstanding contribution to the success of the program. Every worker, every technician, every engineer and manager had the opportunity to personally meet the astronauts. Virgil Grissom's simple words during a speech to the workers at Convair became THE motto through all the Mercury production sites: "Do good work!" Every participant in the program knew very well that anything other than good work could cost the lives of the astronauts.

"Do good work!" could soon be read on countless posters at all the main and secondary contractors. The astronauts spent an increasing

The astronauts during survival training.

The astronauts soon became a tight-knit group.

amount of time in simulators. One of these devices was called MASTIF, an abbreviation of Multiple Axis Space Test Inertia Facility and was the darling of the media. MASTIF was in an abandoned wind tunnel at the Lewis Center and was publicized well beyond its actual value as a training device. The simulator was capable of moving in three rotational axes and two linear axes. It was an arrangement of gimbal-mounted cages that could move freely on a system of concentric rings.

Movement was created by nitrogen gas jets. The device was of considerable size. It had a diameter of twenty-three feet and was suspended from a supporting structure. It could accommodate a two-ton space capsule inside its three Cardan rings and was then capable of spinning the entire combination at up to sixty revolutions per minute. The whole thing took place in hellish noise created by the nitrogen-powered control nozzles.

It was not entirely clear what the device was supposed to simulate. The basic idea behind it was probably to manually regain control of a spaceship whose automatic attitude control had malfunctioned. The device was also used to train the pilot's responsiveness, his ability to multi-task and work under great physical and psychological pressure. The thing was extremely difficult to steer and it turned out that, because of its wild gyrations, even experienced pilots very quickly developed motion sickness.

In February 1960, the first two astronauts, Virgil Grissom and Alan Shepard, arrived to try out the MASTIF. Shepard had to vomit after just a few minutes, but finally the two got used to it and after them all the other astronauts, each of whom completed a week of training with the device by March. The objective was to stop the cockpit in the innermost cage in an upright position, while all the other cages continued turning around them. This could only be achieved by constantly counter-steering with the nitrogen jets. Scott Carpenter once called the training in the MASTIF the ultimate carnival ride.

Much more important than the MASTIF was the second part of the centrifuge program in Johnsville. It began in mid-April 1960, and was carried out with hardware that had been delivered by McDonnell. From then on the astronauts were able to train in the original contour couches, with the original control stick, the original instrument panel and the proper space suits. The interplay of physical stress with the simultaneous monitoring and operation of more than 120 check instruments, electrical switches, circuit breakers and levers, immediately showed the weakness of the current design. The immediate result of the centrifuge tests was the rearrangement of all of the instruments and switches in the capsule. It turned out that in a fully inflated spacesuit at high acceleration or deceleration values, the astronauts could not reach all of the switches and levers. They only succeeded in reaching the instruments from the left center to below left with the left hand and from right center to below right with the right hand. The devices above and in the center could not be reached.

The McDonnell engineers redesigned the instrument panel so that all levers, pushbuttons and switches were arranged in a U shape to the side and below, while only monitored instruments were placed in the other positions. It also turned out that the buttons and levers themselves had to be specially designed so that they could be operated while wearing the space suit. They all had to operate by being slid or pressed. Pulling was difficult or impossible. The pushbuttons needed guides, so that the bulky gloves slid into the correct position or did not slide onto another button under high g forces during launch or splashdown. In some cases levers and pull rings were laid out with an operating force of fifty-five pounds to prevent inadvertent activation. And so it went on.

ATLAS AND LITTLE JOE FAIL

McDonnell production capsule number one arrived at Wallops Island on April 1, 1960. It was a rather naked version lacking most subsystems. It was prepared for a test of the escape rocket, the parachute and the landing system. The rescue system was supposed to be tested on the launch pad, and in space travel history the attempt went down as the "beach abort."

Wallops' personnel prepared the launch rail, which pointed out to sea, the McDonnell people and personnel of the Space Task Group carried out the final tests. Finally they balanced the capsule. For this test the Mercury was not mounted on a Little Joe. Instead the capsule and its escape tower were placed on the launch pylon alone.

On May 9, 1960, everything was ready. The test manager pushed the ignition button and with a roar the escape tower pulled the capsule into the sky. The flight lasted precisely seventy-six seconds from ignition until the perfect parachute deployment. In that time the capsule covered a distance of 2,625 feet and reached a height of 2,428 feet. A Marine Corps helicopter picked it up and half an hour later it was back on Wallops Island. The short test had gone wonderfully; the only minor irregularity was how close the burnt-out escape tower had fallen past the capsule into the water.

Mercury seemed to be developing gradually. This gave the Americans hope. But on May 15, 1960, the Soviets answered back. In the only (publicly announced) Soviet space launch in the first half of 1960, they placed a spacecraft weighing more than 4.5-tons into orbit. The satellite's name did not bode well for the Americans: Korabl Sputnik 1 (Cosmic Ship No. 1). The attempted landing four days later was a failure, because the attitude control system failed and the spacecraft went into a higher orbit instead of slowing for reentry. Possibly, hoped the NASA managers, their people had the better attitude control system. But this beginning of smugness died, when on August 15, 1960, Korabl Sputnik 2 was launched. The next day it brought its two passengers, the dogs Belka and Strelka, safely back to earth.

Nevertheless, by mid-1960 all signs indicated that they were nearing the operational phase of the program. One of these signs was the transfer of Kurt Debus and his team to NASA. The space agency now finally had its own expertise in launch vehicles, launch procedures and launch logistics. Another indication was that Launch Pad 56 for the Redstone and Launch Pad 14 for the Atlas had now been permanently and exclusively assigned to the Mercury program and Hangar S completely belonged to the Space Task Group.

Despite these advances, however, the program was still far from a breakthrough. In this phase this was less due to the capsule than to the launch vehicles, which were still not "man rated." July 29, 1960, when operational qualification of the Atlas

Mission Data	
Mission Name	Beach Abort
Date	May 9, 1960
Launch Site	Wallops Island
Launch Vehicle	Little Joe
Spacecraft	Mercury Number 1
Spacecraft Weight	2,227 lbs
Flight Path	Suborbital
Maximum velocity	186 mph
Flight duration	1 minute, 16 seconds
Flight distance	2,625 feet
Flight path apex	2,427 feet
Landing site	western Atlantic
Recovered by	Marine helicopter

was supposed to take place, was one of the blackest days of the program. That spring, there had been lengthy discussions about which configuration should be used to carry out MA-1. It was originally assumed that it should be a full qualification mission for both the capsule and the launch vehicle.

It was therefore planned to launch the capsule with an active rescue system. The Convair people in particular wanted this, for only with the escape tower atop the Atlas could the real flexing moment and the reactions of the flight control system be tested realistically. Yet sometime during the first half of the year the mission was trimmed and it was finally decided to fly just the naked capsule without the escape tower, so that the weight saved could allow more instrumentation and observation equipment to be carried.

The 1,102-pound escape tower was not the only thing missing, however. Inside the capsule there was also no life support system, no contour couch, no instrument panel and no attitude control jets. A plasticine mass of similar weight replaced the solid fuel in the retrorocket pack.

The primary objective of the MA-1 mission was to test the capsule's structural integrity under the toughest possible reentry conditions it might encounter. To achieve this, the launch vehicle had to reach a velocity of just under 13,047 miles per hour. The apex of the suborbital parabolic track was to be just under ninety-nine miles, flight distance 1,553 miles and highest pressure load 16.3 g. Estimated flight duration was sixteen minutes.

Immediately after McDonnell capsule Number 4 arrived at the Space Task Group in Langley, the technicians installed additional equipment weighing almost 353 pounds. It included two camera systems, two tape recorders and a sixteen-channel telemetry system. There was also extensive instrumentation.

The fully equipped capsule arrived at the Cape on 23 May. All that was missing was the flight instrumentation, the parachute system and the pyrotechnic installations. For the integrated system

The so-called "Beach Abort."

tests, the capsule was moved into the brand-new clean room, which had been installed in Hangar S. After all the usual minor deviations from nominal state had been corrected, the space vehicle was moved to Launch Complex 14, in order to marry the capsule and the rocket, as the technicians vividly put it.

There were problems with the cooling lines and several electrical contacts, however, and the capsule had to be returned to Hangar S. On July 13, it returned to the launch pad, where the technicians carried out the so-called Flight Acceptance Composite Test or FACT. This was completed satisfactorily and on July 21 the next test step, the so-called Flight Readiness Firing, could be carried out. This was a test ignition of the Atlas rocket's engines lasting several seconds, during which the rocket was held fast on the launch pad by clamps. From the capsule the vibration level and acoustic loads were measured and the telemetry transmission and recording systems were tested.

After Flight Readiness Firing, the capsule was again returned to Hangar S. The tape recorder and cameras were removed, reloaded and then reinstalled. The capsule was balanced again, weighed, optically measured and then attached to the nose of the Atlas. Finally capsule number four and Atlas D number 50 were ready for launch. It was July 26, 1960, and the launch was supposed to take place in three days.

In the early hours of July 29, rain pelted the Cape, however, as day began it gradually cleared off. During the last thirty-five minutes of the countdown there were delays totaling forty-eight minutes caused by weather, delays in fueling and difficulties in the telemetry receivers. In the blockhouse, just a few hundred yards from the launch pad, the NASA, air force and Convair launch teams, along with NASA administrator Robert Gilruth, feverishly awaited the liftoff. At 0913 it was time. The Atlas's two steering engines came to life, followed seconds later by the three Rocketdyne main engines. At first the noise was modest, but it grew rapidly and finally the Atlas rose on its jet of flame, fighting

Atlas Number 50, the launch vehicle for mission MA-1 carrying production capsule Number 4, exploded just one minute after liftoff.

gravity as it rose, and seconds later disappeared into the overcast.

The rocket broke through the clouds and was quickly out of sight, but the roaring and crackling of its engines could still be easily heard. Everything appeared to be going normally, but then disaster struck. Contact with the rocket was suddenly lost one minute after liftoff. The telemetry disappeared almost abruptly, but in the previous fractions of a second reported that the pressure differential between the oxygen and oxidizer tanks had fallen to zero.

The failure occurred at the most critical moment of the launch, at the moment of highest dynamic pressure, fifty-nine seconds after leaving the pad at a height of about six miles and a speed of 932 miles per hour. The Mercury telemetry functioned

until the capsule hit the water. At that point none of the metal elements of the outer skin had separated from the airframe. Impact occurred seven miles off the coast of Cape Canaveral. The water there was only thirty-three feet deep, allowing the shattered remains of the capsule to be recovered.

The error could never be identified with certainty. The search was complicated by the fact that the overcast had prevented any optical observation data to be collected, which could have supported the telemetry data that was received. With some probability the cause of the crash was the line leading to the liquid oxygen tank's pressure regulator valve.

Just three weeks after the catastrophic end of Mercury Atlas 1, the Soviets placed the two dogs Strelka and Belka into orbit and twenty-four hours later brought them safely back to earth. It was the first time that a living being had returned alive from orbit.

November 8, 1960, was Election Day in the USA. Kennedy or Nixon was the question in Washington, DC. At Wallops Island, however, they were hoping they would be able to test the escape system in combination with the Mercury production capsule under maximum dynamic load. At 0918, the booster with Mercury Number 3 on top thundered into the sky over Wallops, carrying the hopes and dreams of the engineers of the Space Task Group and McDonnell. These hopes went up in smoke immediately after the rocket left the launch pad, however. Exactly sixteen seconds after the start of the mission, the escape tower fired prematurely.

And then something even worse happened. They could perhaps have salvaged something from the situation and at least carried out a test of the rescue system. But it was not to be. The adapter ring, the capsule and the rocket failed to separate. The Mercury and the burnt-out escape tower remained on the Little Joe. The unfortunate grouping reached an altitude of ten miles and a speed of 1,740 miles per hour, until two minutes and twenty-two seconds after leaving the pad it crashed into the water fourteen miles off the coast of Wallops Island.

McDonnell production capsule Number 4 after recovery from ten meters of water.

Mission Data	
Mission Name	MA-1
Date	July 29, 1960
Launch Site	Cape Canaveral, Launch Complex 14
Launch Vehicle	Atlas D (Number 50D)
Spacecraft	Mercury Number 4
Spacecraft Weight	2,535 lbs
Flight Path	Suborbital
Maximum velocity	1,703 mph
Flight duration	3 minutes, 18 seconds
Flight distance	6.2 miles
Flight path apex	8 miles
Landing site	western Atlantic
Recovered by	Marine helicopter

Resignation was widespread. No one knew what had caused this latest failure. The water at the crash site was about sixty-five feet deep. Recovery divers were able to recover sixty percent of the wreckage from the booster and forty percent of the sad remains of capsule 3. The fragments and the problem were handed over to the Naval Ordnance Test Station in Inyokern, California. There they began an extensive series of tests on the rocket sleds to solve the mystery.

Had the engineers and technicians believed that they had reached the absolute low point of the Mercury program on that November 8, 1960, they had no idea what lay just two weeks in the future.

MR-1 FAILS – MR-1A SUCCEEDS

The first manned Mercury-Redstone flight, MR-3, was supposed to be preceded by the unmanned missions MR-1 and MR-2. Production capsules 2, 5 and 7 were allocated to these three launches.

Capsule Number 2 had been delivered to Cape Canaveral on July 22, and its Redstone launch vehicle followed two days later. In Hangar S, the NASA and McDonnell checkout crews worked around the clock to prepare them for launch. Flight director Christopher Kraft agreed with the launch team under Kurt Debus that the capsule should be flown with the safety system in the open loop mode. This meant that while the capsule would report via telemetry if conditions existed that would require a flight abort, it would not actually carry out the abort. There was great fear of causing a so-called nuisance abort, an unnecessary abort because possibly the tolerances had been set too fine. On this mission, qualification of the launch vehicle was seen as the more important factor.

After several delays, the launch was set for November 7 and the countdown went smoothly until twenty-two minutes prior to launch. Then, however, the launch team was forced to stop the countdown

Mission Data	
Mission Name	Little Joe 5
Date	November 8, 1960
Launch Site	Wallops Island
Launch Vehicle	Little Joe
Spacecraft	Mercury Number 3
Spacecraft Weight	2,513 lbs
Flight Path	Suborbital
Maximum velocity	1,783 mph
Flight duration	2 minutes, 22 seconds
Flight distance	14 miles
Flight path apex	10 miles
Landing site	western Atlantic
Recovered by	Marine helicopter

because they had discovered a leak in the capsule's helium tank, which was to replace the capsule's hydrogen peroxide under operating pressure. The repair, the subsequent tests and resumption of the countdown took two weeks.

At precisely 0900 on the morning of November 21, the countdown for Mercury Redstone 1 reached zero. The tension in the brand-new Mercury control center, which was seeing its first action during this mission, was at the bursting point. What followed has gone down in the history of the Mercury program as "the day we launched the escape tower." Initially the Redstone's engine came to life with a roar. The rocket rose several inches, then the Rocketdyne A-7 engine suddenly shut down, the rocket sat back down on its four fins and came to rest on the launch pad, swaying like a drunk. Everyone waited in breathless silence for what must follow: the explosion of the rocket, followed by a fiery inferno.

But as if the situation were not bizarre enough already, surprisingly there was no explosion. Instead the escape tower—by itself, without the capsule—thundered away, rose to a height of 4,265 feet and crashed onto the beach 1,312 feet from the rocket. Three seconds after the escape tower emerged from the dust, the drogue chute popped from the nose of the Mercury capsule, followed by the main parachute and then the reserve parachute. All three parachutes elegantly wrapped themselves around the body of the still swaying Redstone.

The situation was absurd to such a degree that for a minute no sound broke the silence in the control room. Every one of the witnesses felt as if caught in a nightmare. And yet it was all too real. The bizarre mission of the Mercury Redstone 1 was the most embarrassing failure of the entire Mercury program and its critics wallowed in this incident with enthusiasm.

For a full day no one dared approach the rocket. It was still live. All of the pyrotechnic systems, including the self-destruct system, were live and receiving power. The rocket was not grounded and was filled to the top with fuel. As well, all three parachutes were hanging down the body of the rocket, which was standing free on the pad with no gantry, which could have served as a windbreak. It was possible that at any moment the wind might blow into the parachutes and tip over the rocket. In this extremely dangerous situation McDonnell technician Walter Burke and several colleagues volunteered to disarm the pyrotechnics and reconnect the rocket to the supply lines.

What had caused this bizarre incident? The investigation was difficult but successful. The trigger for the fatal burnout of the Redstone's engine immediately after ignition had been caused by two plugs that had separated from the rocket in the wrong order. One of the two plugs belonged to a control cable, over which activation signals were transmitted to the rocket prior to launch. The other plug belonged to the power cable, which provided electrical power and grounding. Both plugs were attached to the bottom side of one of the tail fins.

At liftoff both were supposed to be pulled out, the control cable plug first, then the power plug. For this launch, however, they had just one control cable available, as used in the military Redstone. For the Mercury-Redstone, with its lesser acceleration, a shorter cable would have been necessary to maintain the correct plug separation sequence. This problem had been recognized prior to launch, and they had therefore used a clamp to shorten the control cable. When the Redstone launched, however, this clamp let go and the cable returned to its original length. As a result of this, the control cable plug did not separate until twenty-nine milliseconds after separation of the power cable, reversing the proper separation sequence.

Because of a brief absence of grounding, however, this tiny time interval was sufficient to allow a power impulse to run through a relay whose job it was to shut down the engine after achieving nominal mission velocity. After the rocket received the signal to shut down the engine, it sent the signal "normal burnout" to the capsule. With this information the capsule initiated two actions, which followed the normal mission sequence: the escape tower, which was no longer needed, separated itself from the Redstone, which had supposedly burned out. In the case of the MR-1, the escape tower separated from the capsule exactly as envisaged. The capsule, however, remained on the rocket, for it was programmed to delay separation until acceleration of the rocket had reached zero. The point of this waiting period was to prevent a separated capsule from possibly being overtaken and struck by a rocket moving on residual thrust. The rocket would only achieve a zero acceleration value when it was in freefall. MR-1 was not in freefall, instead it was still sitting on the ground. The capsule's acceleration sensors correctly detected that there was a constant acceleration of one g, which meant that the capsule must not be released.

Ignition of the escape tower activated the capsule's parachute system. After the air pressure sensors determined that the capsule's altitude was less than 9,842 feet, an accelerated version of the normal parachute sequence was carried out. First

Mission Data	
Mission Name	MR-1
Date	November 21, 1960
Launch Site	Cape Canaveral, Launch Complex 5
Launch Vehicle	Redstone (MR 1)
Spacecraft	Mercury Number 2
Spacecraft Weight	2,712 lbs
Flight Path	Suborbital
Maximum velocity	0 mph
Flight duration	2 seconds
Flight path apex	0.01 feet
Flight path apex	10 miles
Landing site	western Atlantic
Recovered by	Marine helicopter

McDonnell production capsule Number 2, which
was used for both missions MR-1 and MR-1A, is
now on display at NASA's Ames Research Center
in Mountain View, California.

the drogue chute was ejected, followed immediately by the main parachute. Another sensor determined that the main chute was not bearing any weight, therefore assumed that it had not opened, and then also ejected the reserve parachute. And because the Redstone's automatic flight abort system was running in open loop mode, the shutting down of the engine did not initiate the rescue sequence. As envisaged in the open loop scenario, however, the system did report an abort situation and recorded it on the magnetic tape.

The Redstone had been slightly damaged, but it could definitely have been repaired. For a time it was stored as a reserve launch vehicle, but it was never flown and today is on display at the Marshall Space Flight Center. The plug connections were redesigned for all subsequent Redstones. The plug for the power supply was given an extra-long cable, which did not separate from the rocket until after a clear delay, after all the other connections had separated.

Within a few days of the disaster a new MR-1 mission was scheduled: MR-1A. It used the Redstone that had been reserved for MR-3, the first manned mission. The capsule from the unlucky MR-1 mission, Mercury number 2, had survived the disaster virtually undamaged and, with its parachutes repacked, could be used again. The new escape tower came from Mercury number 8 and the antenna cover from space vehicle number 10. Launch date was set for December 19, 1960. That day strong high-altitude winds over Cape Canaveral forced a delay of forty minutes. Then, at the last minute, a magnetic valve in the capsule's hydrogen-peroxide system had to be replaced, which cost another hour. And so it was 11:45 in the morning before the dramatic last ten seconds of the countdown were counted off.

This time everything worked perfectly. The Redstone's A-7 engine operated for 143 seconds. When its fuel supply ran out, speed was 4,846 miles per hour. Propelled by this thrust, the rocket flew on its parabolic course to a height of 130 miles, where the booster and spacecraft separated as

November 21, 1960 went down in the history of the Mercury Program as "the day we launched the escape tower."

planned. The attitude control system functioned perfectly and put the capsule into precisely the right reentry position. Finally it came down on its parachute 234 miles off the coast, almost eighteen miles further than planned.

The pilot of a P2V patrol aircraft was able to spot the capsule from an altitude of 4,265 feet. Thirty-five minutes after it had lifted off, the Mercury capsule was fished out of the water by a Marine helicopter from the aircraft carrier USS *Valley Forge*, and forty-eight minutes after leaving the launch pad it was safely stowed on the deck of the carrier.

On account of the overshooting of the target, the previously planned deceleration values were clearly exceeded and reached a maximum of 12.4 g. This figure, which would be close to the g limit for even a highly trained astronaut, gave some cause for concern.

Later, when the film from the onboard camera was developed, the technicians and engineers saw

Mission Data	
Mission Name	MR-1A
Date	December 19, 1960
Launch Site	Cape Canaveral, Launch Complex 5
Launch Vehicle	Redstone
Spacecraft	Mercury Number 2
Spacecraft Weight	2,712 lbs
Flight Path	Suborbital
Maximum velocity	4,909 mph
Flight duration	15 minutes, 45 seconds
Flight distance	235 miles
Flight path apex	130 miles
Landing site	western Atlantic
Recovery ship	USS Valley Forge

HAM'S BIG MOMENT

On January 3, 1961, there was an organizational change in NASA. The Space Task Group became an autonomous NASA division with a new name. From then on it was the Manned Space Flight Center. At the same time as the change of name, the technical outlook for Project Mercury had also significantly improved. The number one priority now was to ensure that the Mercury-Redstone combination was ready for manned suborbital flights. The decisive test was to be made by Mercury-Redstone 2. And this time there would be a passenger on board.

Redstone booster number 2 and capsule number 5 were selected for this mission. The Mercury had arrived at the Cape on October 11, where in the following weeks it was prepared and tested in Hangar S. The Redstone was flown in from Huntsville, where it had completed a hot run test, on December 20.

Based on the experience with MR-1A, the mission planners decided on a flatter flight path for this mission, with a ceiling of 115 miles instead of 130 and a flying distance of 289 miles instead of the just under 239 miles of the MR-1A.

Capsule number 5 incorporated a series of significant changes compared to its still incompletely equipped predecessors. It was the first Mercury with a space attitude control system, active retrorockets, an environmental control system, radio equipment and the closed loop abort sensing system and also the first flying unit with a pneumatic landing bag.

The sinfully expensive beryllium heat shield was also used for the Redstone flights—not because they needed it, but simply because it was on hand. It had been ordered, manufactured and delivered as a fallback solution in case the ablative shield did not work. After the successful test of the ablative shield on the Big Joe, however, they had decided not to use it for the orbital missions.

As expected, the capsule passenger enjoyed great attention from the media. This passenger was

countless washers, rivets and bits of wire floating about the cabin. This was a clear indication that cleanliness standards had to be raised considerably. Otherwise the flight of the MR-1A was just the right Christmas present for the Mercury team. Keith Glennan called the mission an "unqualified success."

On December 1, Korabl Sputnik 3 was fired into orbit with the dogs Pchyolka and Mushka on board. They were supposed to be recovered the next day, but the maneuver failed and the two dogs lost their lives. Not until decades later did it become known that the Soviets had made another attempt with a Korabl Sputnik on December 22. This time the third stage of the booster rocket failed, but the Soviet rescue system worked and the two dogs Kometa and Shutka were recovered unhurt in Siberia.

The message from this Soviet failure was mixed. The fact that the Soviets had undertaken orbital test flights with a manned system in rapid succession caused a queasy feeling among the Mercury people. They had the almost certain knowledge that they probably wouldn't have a chance of undertaking a manned orbital flight before the Soviets. They did, however, believe they had a good chance of at least being able to carry out a suborbital flight before the Soviets.

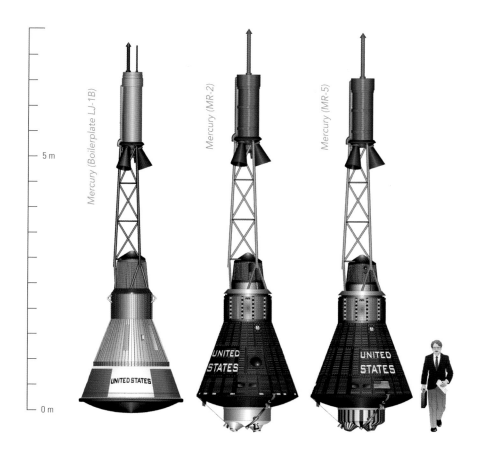

Mercury (Boilerplate LJ-1B)

Mercury (MR-2)

Mercury (MR-5)

5 m

0 m

This drawing shows the differences in external shape between the boilerplate used in the Little Joe 1B experimental mission and the Mercury production capsules Number 5 and Number 9, which were flown on the Mercury-Redstone 2 and Mercury-Atlas 5 missions (with the chimpanzees Ham and Enos).

a three-and-a-half-year-old chimpanzee. He was one of the so-called "astro-chimps," a colony of four female and two male apes, which had been trained by animal trainers at Holloman Air Force Base in New Mexico. Together with their keepers, the chimpanzees had been housed in two barracks behind Hangar S since 2 January. As a precautionary measure against possible contagion, they were divided into two separate groups.

A mockup of the Mercury capsule had been constructed in each of the two shelters, in which the chimps prepared for their tasks. By the end of January each of them was an expert in the art of pressing levers. In order to condition them to perform their tasks correctly, for each correct action they received a banana chip, as opposed to a mild electric shock for each incorrect action.

The day before the launch the veterinarians from Holloman selected two candidates for the flight. They were Chang and Minnie, a male and a female. The ape that would make the flight was to be baptized with the name Ham, in honor of the Holloman Aerospace Medical Center, which looked after him and his colleagues. Nineteen hours before the launch, they were both examined, fitted with

65

Ham takes off on his suborbital spaceflight.

biosensors and placed on a low-residue diet. Four hours before launch they were strapped into their contour couches and taken to the transfer bus. The vehicle arrived at the rocket and there, one-and-a-half hours before launch, Chang—from then on Ham—was selected for the mission.

On that January 31, 1961, preparations at Complex 56 had been underway for hours without problems. Although the weather did not look particularly good and almost six-foot-high waves were reported from the target area, at 0725 the last part of the countdown began. Ham was sealed inside the capsule at 0753 and the acoustic signal to clear the launch pad rang out.

One minute after Ham was put on board, the temperature of one of the inverters began to rise above the allowable limit. The control center cut the power and the device began to cool. The countdown was resumed at 1045. Then there was a series of other minor problems: the elevator to the gantry became stuck, the check of the environmental control system took twenty minutes longer than planned, and the plug cover flaps in the tail of the Redstone jammed. Finally, another cooling period for the inverter had to be initiated, which lasted until 1140.

As the delays went on, Flight Director Christopher Kraft asked the veterinarians if Ham could endure the waiting periods. The animal doctors checked the data from the sensors and confirmed that Ham was fine and that the temperature in the primate couch was a comfortable twenty degrees. Finally the countdown was resumed.

Ignition of the engine took place at five minutes before twelve, and for the first minute everything went exactly according to plan. But then things began to get out of hand. The computer in the control center found that the ascent parabola was beginning to become steeper. Two minutes after liftoff it was so steep that a force of 17 g was calculated for the reentry.

One hundred thirty-seven seconds after ignition the oxidizer was used up and the engine shut down, even before it received the end of burn command.

The Redstone's abort system, which was working in closed loop mode, reported this as a dangerous anomaly and fired the escape tower, which dragged the capsule behind it, and the additional thrust boosted speed by 186 miles per hour. Immediately afterwards the retrorocket pack was jettisoned, which meant that the capsule could not create a braking impulse and thus covered an even greater flying distance.

An equally unexpected but ultimate test of the primate couch happened, when two minutes and eighteen seconds into the flight the cabin pressure abruptly dropped from 380 millibars to 70 millibars. This malfunction was later traced to a defect in the pressure relief valve. During flight in a vacuum, this valve was supposed to be held in the closed position by a spring under tension, which was secured by a locking pin. In the heavy vibration that accompanied liftoff, however, the pin had shaken out, whereupon the valve opened. Normally, this should not have happened until after the main parachute had opened, during descent to the surface of the water. Ham was unaffected, however. The pressure and temperature in his small cabin remained in the normal zone, although the open valve did cause some problems after splashdown.

Because of the over-acceleration caused by the excessively steep angle of flight plus the additional energy produced by the escape rockets, the capsule reached a velocity of 5,840 mph instead of the planned 4,411 mph. Instead of the parabolic flight path's planned apogee of 115 miles, Ham reached an altitude of 157 miles. Instead of four minutes and fifty-four seconds, the period of weightlessness lasted six minutes and twenty-six seconds. Ham splashed down in the Atlantic after a flight lasting sixteen minutes and thirty seconds (fourteen minutes and twenty-five seconds had been planned), 423 miles downrange. He had overshot the target by 130 miles. Maximum pressure during the reentry phase had been 14.7 g. That was 3 g more than planned.

And how had Ham handled all this? In front of him he had a simple instrument panel with two lights and two levers. A force of about one kilogram was required to move the levers. The right lever had to be pressed for a full fifteen seconds. If he forgot, a white lamp lit up to remind him. If he still forgot, he received a mild electric shock. To make the exercise more psychologically challenging, there was a second blue lamp. If this lit up, which it did roughly every two minutes, then he had to operate the left lever within five seconds to avoid being reminded of his job by an electric shock on the soles of his feet. The system was in operation from liftoff until the parachute opened. Ham carried out his tasks almost perfectly. He pressed the right lever about fifty times and received two shocks for bad timing. He was perfect with the discrete avoidance lever and did not have a single failed attempt. His average reaction time for operating the second lever was 0.82 seconds, almost exactly the same figure he had achieved on the ground. As well, the monitoring camera in the capsule again showed an incredible amount of dust and small parts flying about. The cleanliness problem had obviously remained unsolved.

When Ham's capsule splashed down at 1212, there was no one in sight far and wide. Twelve minutes later the recovery forces received the first radio signal from the capsule. Triangulation revealed that it had come down almost sixty-two miles away from the nearest recovery ship, the destroyer *Ellison*. Twenty-seven minutes after splashdown, the capsule was sighted by the crew of a P2V patrol aircraft floating upright in the water. Mercury control calculated that it would be at least two and a half hours before the Ellison reached that point and therefore asked the navy to dispatch helicopters. The closest ship with a helicopter on board was the landing vessel *USS Donner.*

By the time the helicopter reached the capsule the picture had clearly changed. The capsule was now on its side, obviously taking on water and in danger of sinking. The high waves were punishing it badly. The pneumatic landing bag had been torn, the beryllium heat shield attached to it had twisted on the landing bag and was now banging against

the capsule in rhythm with the waves, and had already knocked two holes in the pressure cabin's titanium shell. Prior to recovery the shield separated completely and sank. It was discovered that the water was entering the capsule through the open air inlet. When the two helicopter pilots, Lieutenants John Hellriegel and George Cox, finally hooked the capsule, they estimated that there was about eighty-eight gallons of water on board. Inside they found the unflappable Ham. After a rough flight back to the *Donner*, Hellriegel and Cox lowered the capsule onto the deck and minutes later Ham was free. He was in good condition and soon ate an apple and half an orange.

Ham's mission had given the program managers the confidence that a man could now complete the flight even if dangerous defects arose. Even more: some of the problems were of a type that a man could resolve during the mission. Apart from the landing bag, which had to be reworked, from NASA's point of view the Mercury system was operational.

This confidence was not shared by the people around Wernher von Braun, however. Twice the booster rocket had not fully lived up to expectations. On both flights, MR-1A and MR-2, there had been

Mission Data	
Mission Name	MR-2
Date	January 31, 1961
Launch Site	Cape Canaveral, Launch Complex 5
Launch Vehicle	Redstone
Spacecraft	Mercury Number 5
Crew	Chimpanzee Ham
Spacecraft Weight	2,667 lbs
Flight Path	Suborbital
Maximum velocity	5,841 mph
Flight duration	16 minutes, 39 seconds
Flight distance	421 miles
Flight path apex	157 miles
Landing site	western Atlantic
Recovery ship	USS Donner

an inexplicable over-performance of the booster. Von Braun, Küttner and Debus therefore requested another qualification mission for "old reliable." In their opinion the rocket was not as reliable as its name suggested.

After his spaceflight, Ham lived for another seventeen years in the zoo in Washington, DC. In 1981, he went to an open-air enclosure in North Carolina, where, on January 19, 1983, he succumbed to an infection. Minnie, his "replacement wife," had nine offspring. When she died at the age of forty-one on March 14, 1994, she was the last surviving astro-chimp.

Ham on the USS Donner immediately after recovery.

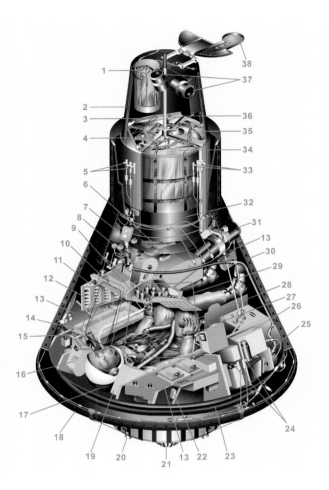

1. Braking parachute container. 2. Antenna cone. 3. UHF antennas for ascent and recovery. 4. HF mast antenna. 5. Yaw control jets. 6. Voice and telemetry data recorder. 7. Nitrogen tank. 8. Observation camera for astronaut. 9. Mission abort lever. 10. Main instrument panel. 11. Left instrument panel. 12. Safety switch. 13. Instrument package. 14. Observation window. 15. Survival equipment. 16. Observation camera for instrument panel. 17. Individual astronaut couch. 18. Ablative heat shield. 19. Water tank. 20. Solid-fuel separation rockets (3) for separating from the launch vehicle. 21. Solid-fuel braking rockets (3). 22. Radiometer. 23. Pressurized gas container for attitude control. 24. Roll control jets. 25. Part of the life support system. 26. Communications equipment. 27. Radiometer. 28. Control stick for three-axis attitude control. 29. Entry hatch. 30. Double-wall pressure cabin. 31. Periscope. 32. Airbags for ejecting parachute. 33. Pitch control jets. 34. Reserve parachute. 35. Main parachute. 36. Recovery signal light. 37. Horizon seeker. 38. Destabilizing flap (correction of reentry direction).

THE MERCURY CAPSULE

Altogether, twenty Mercury capsules and an unknown number of boilerplates were made. No two capsules were exactly the same; all were tailor-made for their respective missions. On almost every mission, therefore, there were minor deviations from the information contained in the capsule specifications offered here.

The pilot in his space suit accounted for fifty-one miles of the total orbital mass of the capsule.

The Mercury's overall length was calculated from the lowest point of the retro-system to the aerodynamic tip of the escape tower.

On a normal mission, the escape tower was ignited after reaching a safe height, in order to separate it from the capsule. In the event that it was used (during an aborted launch) there was a smaller separation rocket that separated the escape tower from the capsule.

The booster separation rockets served to distance the capsule from the booster rocket after separation. The retro-rockets slowed the spacecraft's orbital speed by about 218 miles per hour and thus enabled it to return to earth.

The maximum temperature of the heat shield was 3,000 degrees Fahrenheit. This was achieved

Mission Data	
Manufacturer	McDonnell
First flight	September 9, 1959
First manned flight	May 5, 1961 (MR-3)
Last flight	May 15, 1963 (MA-9)
Number of flights	16
Number suborbital	8
Manned flights suborbital	2
Manned flights orbital	4
Maximum mission duration	1.5 days

Specification	
Capsule length	13 feet
Length including escape tower	26 feet
Diameter at the base	6.2 feet
Usable volume	5.58 square feet
Launch weight including escape tower	4,266 lbs
Cabin atmosphere	pure oxygen
Orbital weight including retrorockets	2,987 lbs
Attitude control engines	6 x 105 N (23 lbs); 6 x 4.45 N (1 pound)
Attitude control fuel	hydrogen peroxide

Retro System	
Height	19.6 inches
Diameter	3.28 feet
Total weight	523 lbs
Fuel weight	452 lbs
Fuel	solid
Separation rockets	3 x 1.78 kN (52,000 lbs)
Separation rocket burn duration	1 second
Braking rockets	3 x 4.45 kN
Braking rocket burn duration	10 seconds each

Escape Tower	
Manufacturer	Grand Central
Length	17 feet
Total weight	1,279 lbs
Escape rocket thrust	231.3 kN (1 pound)
Burn duration	1 second
Separation rocket thrust	3.56 kN (800 lbs)
Separation rocket burn duration	1.5 seconds
Fuel	solid

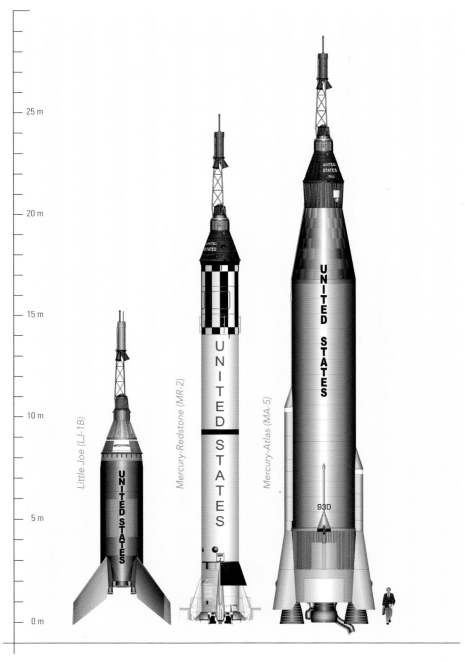

25 m

20 m

15 m

10 m

5 m

0 m

Little Joe (LJ-1B)

Mercury-Redstone (MR-2)

Mercury-Atlas (MA-5)

UNITED STATES

UNITED STATES

UNITED STATES

93D

MR-2

Launch Vehicles

Atlas-D (Mercury-Atlas)

Manufacturer	Convair
Launch site	Cape Canaveral, Launch Complex 14
Overall length	94 feet
Diameter at the base	16 feet
Launch weight	120 tons
Launch thrust	1,586 kN (356,000 lbs)
Payload (orbital)	3,031 lbs

Atlas D Booster

Engine	2 x LR-89-NA-5
Thrust at sea level	2 x 662 kN (149,000 lbs)
Thrust in a vacuum	2 x 758 kN (170,400 lbs)
Burn duration	135 seconds
Fuel	Kerosene
Oxidizer	Liquid oxygen
Weight	6,742 lbs
Length	15 feet
Diameter	16 feet

Atlas D Sustainer

Engine	1 x LR-105-NA5
Thrust at sea level	262 kN (59,000 lbs)
Thrust in a vacuum	363 kN (81,600 lbs)
Fuel	kerosene
Oxidizer	Liquid oxygen
Empty weight (fueled)	16,094 lbs (320,993 lbs)
Length	66 feet
Diameter	10 feet

Operational History

First Launch	April 14, 1959
Mercury launches (successful)	8 (6)
Last launch	May 15, 1963

Mercury-Redstone

Manufacturer	Chrysler
Launch site	Cape Canaveral, Launch Complex 5
Overall length	83 feet
Diameter	6 feet
Span fins	14 feet
Launch weight	30 tons
Launch thrust (thrust in a vacuum)	350 kN (414 kN) 78,700 lbs (93,000 lbs)
Payload (suborbital)	4,189 lbs
Engine	1 x Rocketdyne A 7
Burn duration	144 seconds
Fuel	Ethyl alcohol
Oxidizer	Liquid oxygen

Operational History

First Mercury launch	November 21, 1960
Mercury launches (successful)	6 (5)
Last Mercury launch	July 21, 1961

Little Joe 1

Manufacturer	North American
Launch site	Wallops Island
Overall length	49 feet
Diameter	6.5 feet
Span fins	21 feet
Launch weight	20 tons
Payload (suborbital)	3,086 lbs

Booster

Recruit	4 x
Fuel	Solid
Length x diameter	9 feet x 9 inches
Thrust	each 167 kN (37,500 lbs)
Burn duration	1.5 seconds

Sustainer

Pollux or Castor	2 x or 4 x
Fuel	Solid
Length x diameter	20 feet x 2.5 feet meters
Thrust	each 259 kN (58,200 lbs)
Burn duration	37 seconds

Operational History

First launch (test LJ 1)	August 21, 1959
Total launches (successful)	7 (5)
Last launch	April 28, 1961

at a speed of 14,911 miles per hour at an altitude of twenty-five miles.

The drogue chute, with a diameter of six feet, was deployed at an altitude of between four and five miles. The main parachute, with a diameter of sixty-three feet, was deployed at a height of about three miles; it was initially reefed and deployed fully at about two miles. The heat shield was detached shortly before the capsule landed in the water. The four-foot-long pneumatic landing bag, a perforated, rubberized glass fiber skirt, was attached to the shield. This device softened the impact of touchdown. As a rule, the main parachute was automatically separated from the capsule.

ONCE QUALIFIED AND ONCE NOT

While the Redstone was getting closer to qualification for manned missions as the year 1960 gave way to 1961, a big question mark still hovered over the Atlas. The cause of the explosion of the MA-1 on July 29 had not been fully explained. It was conjectured that on reaching the zone of maximum dynamic pressure, the structure between the body of the Atlas and the Mercury had collapsed. A reinforcement had been added to this area, but if the cause lay elsewhere it would be useless.

The question now was whether they should carry out the flight without a definitive explanation for the cause of the explosion. The pressing schedule provided the answer: they had to risk it. The launch was scheduled for the morning of February 21, 1961.

The mission was to be flown by Mercury production capsule number 6 and the Atlas with the serial number 67D. The launch vehicle was modified, with a eight-inch-wide steel reinforcement added to the transitional structure between the dome of the liquid oxygen tank and the capsule. The mounting for the oxygen regulator valve, which was also in the dome of the tank, was likewise reinforced and the adaptor between the capsule and booster was also strengthened.

The flight profile of the failed MA-1 mission was to be retained. The objective was to carry out a suborbital flight with a shallow parabolic flight path and a reentry speed of about 12,426 miles per hour. This would create the toughest possible reentry conditions that a Mercury could experience in an abort situation during an orbital launch. The abort sensor array would operate in the closed loop mode, which meant that the system would detect the— artificially produced—loss of the sustainer engine and activate the escape tower.

Launch preparations had proceeded with no major problems, and on that sunny morning of February 21, 1961, the countdown clock ticked toward zero. The weather on the Cape was perfect, but in the splashdown zone 1,243 miles away it was quite cloudy, and the launch had to be postponed for an hour. Then, at 0912, MA-2 finally left the launch pad with a thunderous roar and for the next two minutes no one dared to breathe. An audible sigh of relief went through the control center when the zone of maximum dynamic pressure (Max Q) had been passed. Telemetry showed that the sequence of operations was proceeding like clockwork: burnout of the booster, jettisoning of the booster engines, shutdown of the sustainer engine, activation of the escape tower, separation of the capsule from the launch vehicle, separation of the escape tower from the capsule, turning of the capsule, positioning for ignition of the retro-rockets, firing of the retro-rockets and jettisoning of the expended retro-rocket pack.

The launch of Atlas test mission MA-2.

Mission Data	
Mission Name	MA-2
Date	February 21, 1961
Launch Site	Cape Canaveral, Launch Complex 14
Launch Vehicle	Atlas D (Number 67D)
Spacecraft	Mercury Number 6
Spacecraft Weight	2,557 lbs
Flight Path	Suborbital
Maximum velocity	13,234 mph
Flight duration	13 minutes, 58 seconds
Flight distance	1,429 miles
Flight path apex	113 miles
Landing site	central Atlantic
Recovery ship	USS Donner

Telemetry contact was then lost due to the great distance, but just three minutes later the destroyer *Greene* reported a visual sighting of the capsule and the Atlas booster during reentry. Within ten minutes a bearing was taken on the radio beacon and once again the USS *Donner* dispatched its helicopter. Just twenty-four minutes after splashdown, Mercury capsule number 6 was fished out of the water. Less than an hour after launch it was secured to the deck of the Donner. MA-2 had been a complete success. Virtually everything had run nominally. The problem with the Atlas was solved, capsule and heat shield were in very good condition, even better than they had been on Big Joe, and the transmission of telemetry had functioned perfectly.

Proof had been obtained that the capsule was capable of mastering the hardest possible regular reentry conditions. The world could have been so nice if there had not been a pair of unresolved problems. One of them was the demonstration of a flight abort with a production capsule at the maximum dynamic pressure that could be encountered during the launch of an Atlas.

Production capsule number 14 was used for the test with Little Joe 5A. The version of the Little Joe used had just two Castor solid-fuel rockets, as the test was to take place at an altitude of just six miles at a speed of Mach 1.52. The dynamic pressure under those conditions was equivalent to a force of 4.5 tons per square yard. The Little Joe would thus almost exactly simulate the dynamic pressure profile of the Atlas at that altitude and speed.

The launch took place at 11:49 on March 18, 1961, after a four-hour delay caused by problems working through the countdown checklist and it developed spectacularly and quite differently than planned and hoped. The four Recruit and two Castor rockets roared to life, and the Little Joe thundered from the launch rail angled towards the sea into the sky. For twenty seconds it looked as if everything would go according to plan this time. But then the escape tower again fired prematurely, and once again, as on Little Joe 5, the capsule remained on the booster.

This time, however, the story went further. Thirty-five seconds after launch, at the planned time, the preprogrammed abort signal was issued, causing the clamping ring to release. Because of the aerodynamic pressure, however, it remained on the tip of the rocket until, eight seconds later, following a manual command from the ground station, a single small solid-fuel rocket fired. It had been retrofitted as an emergency separation system. At the same time, another pyro system initiated separation of the tower from the capsule. This emergency separation system was not supposed to have fired until the apogee of the parabolic flight path, where the dynamic pressure was at most only 441 pounds per square yard. When it was initiated, the actual value was ten times as much.

For this reason, the propellant charge did little against the wind pressure. The capsule immediately began to tumble, in the process colliding either with the launch vehicle or the escape tower and tearing away the retro-rocket pack. The aerodynamic forces affecting the capsule, and perhaps also the escape tower when it separated, ruptured the nose cone with the antenna canister and the parachute containers. Both parachutes opened under high g loads, causing several of the lines to break and ultimately tearing many of the parachute ribbons.

The pneumatic landing bag was activated by the barostats at a height of two miles. After the capsule had involuntarily activated both parachutes at high altitude, it drifted away in the wind. Splashdown was envisaged for a point ten miles off the coast, but now the capsule came down eighteen miles out in the Atlantic. To make matters worse, the parachute separation mechanism also failed to work, making it impossible for the helicopter to pick up the capsule.

An hour after landing, a ship picked up the capsule. It was in very good condition despite its rough handling. It was later determined that the deployment loads on the parachutes had been six times as high as normal.

A detailed analysis of the flight showed that LJ-5 and LJ-5A had failed for the same reason. Both boosters showed a structural deformation near the clamping ring that damaged the electrical and mechanical components of the separation system.

The goal of qualification of the capsule under maximum abort conditions had thus once again not been achieved. Therefore the test had to be rescheduled. The date for the test was set for four weeks later and the last available Little Joe had be used. Mercury capsule 14 was cleaned, repaired and fitted with new sensors, instruments, parachutes and telemetry equipment.

Mission Data	
Mission Name	Little Joe 5A
Date	March 18, 1961
Launch Site	Wallops Island
Launch Vehicle	Little Joe
Spacecraft	Mercury Number 14
Spacecraft Weight	2,513 lbs
Flight Path	Suborbital
Maximum velocity	1,783 mph
Flight duration	4 minutes, 25 seconds
Flight distance	18 miles
Flight path apex	8 miles
Landing site	western Atlantic

YURI GAGARIN AND ONE INTERMEDIATE STEP TOO MANY

On March 9, 1961, the USSR announced the launch of its fourth—official—Korabl Sputnik. Once again there was a passenger on board: the female dog Chernushka. On the same day she was brought safely back from orbit.

Four weeks earlier, on February 13, 1961, NASA administrator Robert Gilruth, Max Faget and several other high-ranking NASA officials had met with the project heads of McDonnell and Wernher von Braun's group. The sole item on the agenda was the question of whether they should launch MR-3 with a man on board. Public pressure was great, the astronauts had repeatedly expressed their readiness to fly the mission, and the Soviet Union appeared to have almost completed its preparations.

But the developers of the Redstone did not have as much faith in their own product as the NASA leadership. Wernher von Braun and Kurt Debus stressed that it was necessary to complete at least one fully successful unmanned flight in order to qualify the Redstone for manned missions.

The NASA leadership therefore finally agreed to add another test mission to the Mercury timetable. At the same time the fateful decision was made to postpone MR-3, the first manned mission, until April 25, so that this new mission could be inserted on

March 28. There was little question as to the technical necessity of this decision, and while the potential political consequences were clear, they lived in the hope that they would not be realized.

The question now arose of the Redstone's payload. One of the expensive production capsules was out of the question, even if one had been available. And so the boilerplate from Little Joe 1B of January 1960 was repaired. The mission was dubbed MR-BD, for Mercury-Redstone Booster Development. Recovery of the capsule or the remains of the rocket was not envisaged, and so the MR-BD was left to von Braun's team, without the Manned Space Center having much influence on it. In the first two weeks of March, the Redstone engineers worked on tuning their rocket. The reason for the over-performance on the previous mission was found. It was based on a control valve that had failed to properly regulate the flow of hydrogen-peroxide to a steam generator that in turn drove the fuel pumps. Another problem was harmonic resonances that appeared in the connection between the rocket and capsule. This area had to be worked on, with stiffeners and additional insulation. In order to precisely measure the resonance behavior, sixty-five additional telemetry sensors were distributed over the structure. And finally, and this was the most complicated modification, the burnout behavior was changed to prevent an unnecessary abort signal as on the MA-2. Finally, the whole thing was equipped with a dummy escape rocket.

On the morning of March 24, 1961, the countdown was completed without the slightest problem. At 12:30 the MR-BD lifted off smoothly from Cape Canaveral on its preprogrammed trajectory and flew without problems until engine burnout. Its terminal velocity was still just under sixty-two miles per hour faster than calculated, but this was much closer than on the two previous missions. The entire combination fell into the Atlantic 308 miles downrange and announced its presence by means of a Sofar bomb. Evaluation of the telemetry revealed that the vibration level was now within acceptable limits. Wernher von Braun's team was satisfied, Kurt Debus'

Mission Data	
Mission Name	MR-BD
Date	March 24, 1961
Launch Site	Cape Canaveral, Launch Complex 5
Launch Vehicle	Redstone
Spacecraft	Boilerplate
Spacecraft Weight	2,513 lbs
Flight Path	Suborbital
Maximum velocity	5,120 mph
Flight duration	8 minutes, 23 seconds
Flight distance	307 miles
Flight path apex	113 miles
Landing site	western Atlantic

team was satisfied, and so was the rest of NASA. The Redstone was now considered man-rated, and the next suborbital flight could be manned.

The day after the mission by MR-BD the Soviets announced the successful launch and recovery of their fifth Korabl Sputnik. On board was a dog named Zvyozdochka (Starlet). Thus their last two flights had been successful. If the Soviets were using the same criteria as the Americans—namely two successful test missions in succession—then the next Soviet mission would be manned.

At the beginning of April 1961, Redstone number 7 for the MR-3 flight was placed in a vertical position on the launch pad of Launch Complex 5 at Cape Canaveral. McDonnell production capsule number 5 was envisaged for the program's first manned launch. It was undergoing final checks in Hangar S. In those busy weeks, however, full attention was scarcely given to the execution of MR-3, for two critical unmanned missions were scheduled for the next two weeks, namely the third attempt at the maximum dynamic pressure test for the production capsule and mission MA-3 for orbit qualification.

On April 10, there were several rumors from correspondents in Moscow about an imminent Soviet manned space flight. When this was not confirmed

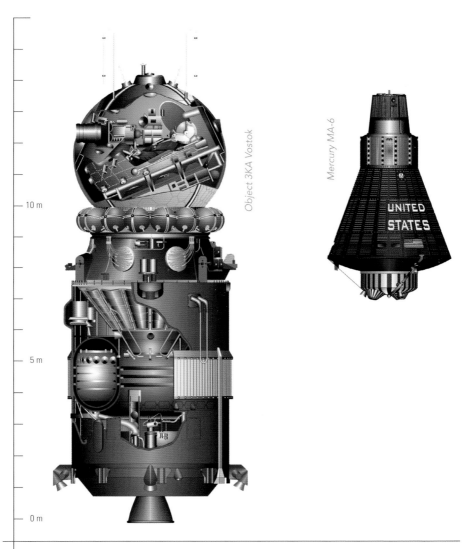

Object 3KA Vostok

Mercury MA-6

UNITED
STATES

10 m

5 m

0 m

This drawing shows the completely different approach that the Soviet Union adopted for its manned Vostok spacecraft. On the left is Vostok 1, in which Yuri Gagarin orbited the earth in April 1961. This space vehicle had three times the mass and was significantly larger than Mercury. On the right is a Mercury capsule based on Friendship 7,

in which John Glenn took off on his orbital flight in February 1962. Note that Vostok 1 is shown here with the third stage of the launch vehicle. The spacecraft itself ends at the silver ring that encircles the cylindrical structure at the level of the graphic bar. At the end of this double cone was the engine of the Vostok's service module.

by the following day, the excitement quickly died away. Thus came April 12, 1961.

The day began in the early morning hours in the USA with a report by the Associated Press, which was based on a press release by TASS, the Soviet news agency. Translated, it read: "Vostok, the first manned spaceship in the world, was launched by the Soviet Union on April 12 on a flight around the world. The first space navigator is Soviet citizen Major Yuri Alexeyevich Gagarin. Bilateral radio communication with Major Gagarin was established and maintained."

It was the moment that the NASA officials had seen coming for months. At 07:45 James Webb spoke to the nation in a radio broadcast and congratulated the Soviets on their accomplishment. Then he expressed NASA's disappointment at having lost the race, and finally assured the American people that Project Mercury would be carried on as planned. He described the process as a logical first step by the USA in the conquest of the moon. It was a step that had to be taken, whether others put a man in earth orbit before the USA or not.

Webb's speech was considered, balanced and reasonable. Nevertheless the news of the Soviet first was a devastating disappointment for the Americans, and the astronauts were probably more disappointed than anyone. They knew how close they had been to becoming first in space. Not in orbit like the Soviets, that was another league and they knew it, but at least in a suborbital flight. For a public with no knowledge of these things, it would have been almost the same thing. If the decision to carry out MR-BD had not been taken and MR-3 had gone ahead straightaway, it might have happened.

The points that manifested themselves to the staff of the Manned Space Center were different. The Vostok capsule was three times as heavy as the Mercury. The Soviets had not stopped with preliminary suborbital missions.

Vostok had a mixed gas atmosphere with sea level air pressure. The Soviet spacecraft had a jettisonable service module. It had a flight path with a high inclination (sixty-five degrees), during most of which it was autonomous and out of radio contact with the ground. All of this gave the impression of an immense lavishly planned and executed program. Mercury, on the other hand, manifested itself as the minimal concept, which it in fact was.

A comparison between Yuri Gagarin's R 7 Vostok rocket and the Redstone rocket that carried Alan Shepard and Virgil Grissom.

Vostok R 7

1. Payload fairing with catapult opening. 2. Vostok cabin with ejection seat. 3. Service module. 4. TDU-1 braking engine. 5. Ring-shaped Vostok stage oxidizer tank. 6. Thruster. 7. 8D719 Vostok stage engine. 8. Open adapter. 9. First stage core oxidizer tank. 10. Strap-on booster attachment. 11. First stage core. 12. First stage strap-on booster. 13. Strap-on booster oxidizer tank. 14. Oxidizer line. 15. First stage core kerosene tank. 16. Strap-on booster fuel tank. 17. Oxidizer line. 18. First stage core hydrogen-peroxide tank. 19. Strap-on booster nitrogen tank. 20. Strap-on booster hydrogen-peroxide tank. 21. First stage core nitrogen tank. 22. Thrust frame. 23. RD-107 first stage strap-on engine (total of 5). 24. Stabilizing fins (total of 4). 25. Thrusters. 26. RD-108 two-stage engine.

Mercury-Redstone 3

27. Escape tower. 28. Framework structure. 29. Antenna cone. 30. Pressure cabin. 31. Porthole. 32. Entry hatch. 33. Separation ring. 34. Adapter. 35. Separation and retrorocket pack. 36. Ballast section. 37. Instrument compartment. 38. Guidance system. 39. Fuel tank (ethylene). 40. Oxidizer tank. 41. Fuel line. 42. Turbopump. 43. Thrust frame. 44. Turbine exhaust nozzle. 45. Jet vane. 46. Rocketdyne A-7 expansion nozzle. 47. Stabilizing fins. 48. Aerodynamic rudder.

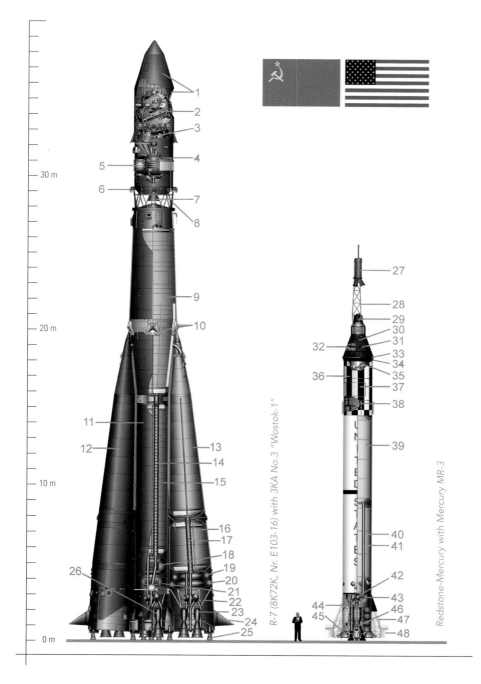

R-7 (8K72K, Nr. E103-16) with 3KA No.3 "Wostok-1"

Redstone-Mercury with Mercury MR-3

SUCCESSFUL FAILURES

On April 10, 1961, the Mercury-Atlas 3 pairing, consisting of the Atlas 100D and the McDonnell production capsule number 8 were sitting on the launch pad waiting to launch. This mission had originally been planned as a longer ballistic flight over Bermuda with splashdown in the Atlantic beyond. One of MA-3's test objectives was to test the collective ability of the control center on the Cape and the tracking station on the Bermudas to carry out a flight abort just prior to achieving orbital speed. And of course the Atlas D was to be qualified as launch vehicle for a manned mission.

After Gagarin's flight, NASA decided that a suborbital mission would not seem ambitious enough for the public. It had now been decided that MA-3 would go into orbit and complete a single orbit of the earth at an altitude of about 100 miles. Technically, that was not that different than what had gone before, but it did require that the landing zone be moved. This was no longer in the South Atlantic, but instead in the Atlantic about 498 miles east of Cape Canaveral.

MA-3 had a passenger, a so-called crewman simulator. It was a machine whose job it was to simulate the metabolism of a man on board. It could inhale and exhale and produced carbon dioxide, heat and water vapor. The device was to be used to test the life-support system.

The launch took place at 11:15 in the morning on April 25, 1961. The rocket failed to follow the envisaged inclination program and veered to the east instead of climbing vertically into the sky. At precisely forty seconds after launch the safety officer blew up the rocket. Seemingly a simple failure, but in fact this incident saw to it that the rocket was subsequently qualified for manned use. And that was because the rescue system functioned perfectly during this unintentional flight abort.

The escape tower pulled the capsule from the rocket, rose to a height of five miles and separated from the capsule. This was followed by a clean parachute sequence with the drogue and main chutes. The recovery team acted exactly as if an astronaut's life was at stake. The capsule floated beneath its parachute, came down in the water several hundred yards from the beach and was immediately recovered. It was virtually undamaged, was overhauled and reused for MA-4, the next mission.

After the intentional explosion, only a few parts of booster 100D were found, including the programmer, which fell on the beach and buried itself six feet deep in the sand. It had been fed a wrong code.

Meanwhile, on Wallops Island the seventh and last Little Joe booster had been fitted with capsule number 14 in order to repeat the unsuccessful flights of LJ-5 and LJ-5A. Rocket and capsule were even more heavily instrumented and equipped with more sensors than before. The checkout was even more precise and careful than before, in order to assure that this time the intentionally-caused flight abort would also happen at the correct time.

It was envisaged that the rocket would follow a steep flight path to a height of nine miles. Then the escape tower would become active while the capsule was under a dynamic pressure of 4.8 tons per square yard, pull the Mercury from the booster and initiate the parachute sequence.

When, finally, at 09:03 on April 28, 1961, Little Joe 5B thundered away from the launch pad, all the neck hairs of the observers stood up when they saw that one of the Castor boosters had ignited five seconds late. As a result, the flight path was much flatter than anticipated and thirty-three seconds after liftoff the combination had only reached a height of three miles. There, where the air was much denser, dynamic pressure was almost ten tons per square yard. The abort itself was initiated at the correct time and all events went as planned.

*Atlas D Number 100 had to be destroyed by the
safety officer after precisely forty seconds of flight.*

Mission Data	
Mission Name	MA-3
Date	April 25, 1961
Launch Site	Cape Canaveral, Launch Complex 14
Launch Vehicle	Atlas D (Number 100)
Spacecraft	Mercury Number 8
Spacecraft Weight	2,602 lbs
Flight Path	Suborbital
Maximum velocity	1,180 mph
Flight duration	7 minutes, 19 seconds
Flight distance	1 mile
Flight path apex	4.5 miles
Landing site	beach at Cape Canaveral

Mission Data	
Mission Name	Little Joe 5b
Date	April 28, 1961
Launch Site	Wallops Island
Launch Vehicle	Little Joe
Spacecraft	Mercury Number 14
Spacecraft Weight	2,601 lbs
Flight Path	Suborbital
Maximum velocity	1,777 mph
Flight duration	5 minutes, 25 seconds
Flight distance	8.5 miles
Flight path apex	8.5 miles
Landing site	western Atlantic

Recovery by helicopter was quick, even though the flat trajectory meant that the capsule had flown two miles farther out to sea.

Despite the booster problem, the test was a complete success. The abort conditions were much harder than those of the planned scenario, and yet everything had functioned perfectly. One of the great obstacles for the manned use of the capsule had been overcome. All of the Mercury program systems, with the exception of the Atlas, were now qualified.

MERCURY BECOMES MANNED

On February 22, 1961, the Space Task Group announced that the astronauts Alan Shepard, John Glenn and Virgil Grissom had been selected to begin mission training for flight MR-3. By then this was nothing new to the three. Just before the New Year, Robert Gilruth had called all seven astronauts together and informed them who would be the first to fly and what the replacement list looked like. If Shepard could not fly for whatever reason, then Grissom would take his place. His replacement in turn was John Glenn.

The decision by the Manned Space Flight Center to train just three astronauts for the Mercury-Redstone mission made sense. This allowed the remaining men to prepare for the support jobs they would be doing in support of MR-3. And they could also begin preparing for the orbital Mercury-Atlas missions.

But Shepard, Grissom and Glenn were, as the astronauts had demanded from the beginning, an integral part of the flight preparations. For example, when the capsule was tested in a vacuum chamber, Shepard sat in the Mercury in his space suit.

Probably the most valuable training devices for the astronauts during this phase were the two procedures trainers made by McDonnell. Later in the program they would also be called Mercury simulators. These devices were very close to a modern flight simulator. One was in Langley, the other at the Cape. In this training device the astronauts lay on their backs in a capsule mockup and trained for a specific mission.

The trainer could be linked to the Mercury Control computer. Together the controllers and astronauts sweated out many joint exercises. During this intensive training period between February and April, Shepard "flew" 120 simulated Mercury-Redstone missions.

The decision to use capsule number seven for this launch had been made almost a year earlier. In principle it could also have been capsules eight, nine or eleven, which were also checked-out and standing by at the Cape; but Capsule 7 was the reference capsule for the manned missions, McDonnell's best product so far. It did, however, have the disadvantage of not having a window, unlike the later capsules. Instead it had two portholes and a periscope.

Redstone number 3 had initially been envisaged for MR-3, but then it was needed for mission MR-1A. As part of this decision, it was decided that Redstone number 7 would be used for the MR-3 mission. The number 7 thus became something of a fateful number for the flight: capsule number 7 on rocket number 7 in the first of 7 planned Mercury flights for the 7 astronauts. When Shepard was asked what call sign he wanted to use for the mission, after lengthy consideration he replied: "Friendship 7."

One of the many training modules in this phase consisted of running through the operational sequence of launch day in greater detail. That began with the pre-launch medical examination, breakfast, putting on the space suit, the transfer from Hangar S to the launch pad and then the launch preparations in the capsule. At the end of each operational sequence there was a simulated flight in conjunction with Mission Control. During both of the first two "practice flights," which took place on April 18-19, 1961, the gantry remained on the rocket and the capsule hatch stayed open. During the simulation on April 20 the hatch was closed and the life-support system activated. Then the gantry was removed and the rocket stood alone on the pad. Training like this and in the procedures trainer continued until two days before the scheduled flight. It was supposed to take place on May 2, 1961.

Not until launch preparations had been concluded was the recovery process with the helicopter defined. It was originally planned that

the helicopter would pick up the capsule with the astronaut inside and then fly to the recovery ship. John Glenn protested against this vigorously. As a former operational pilot with the Marines, it was clear to him that this was much too dangerous, both for the astronauts and the crew of the helicopter. If a problem developed with the empty capsule on the hook, the helicopter pilot could jettison it without qualms. If an astronaut were on board, he would risk everything in order to avoid jettisoning the capsule, thus endangering the astronauts and the crew of the helicopter.

Not until April 15 was the future standard procedure laid down. It looked like this: the helicopter was to hover over the capsule and first determine the precise modalities with the astronaut by radio. The copilot would then cut the antenna (which was necessary to get at the recovery lug), attach the capsule recovery loop and raise the capsule slightly out of the water. By then the astronaut would have unfastened his safety harness and the hatch would be completely out of the water. Then the astronaut would open the hatch, crawl halfway out and grab the "horse collar," a sling that was lowered from the helicopter and resembled a horse collar. The

John Glenn, the second alternate for Alan Shepard on the Mercury-Redstone 3 mission, readies himself in the procedures trainer.

Alan Shepard (left) and John Glenn eating breakfast prior to launch.

helicopter was then supposed to lift the astronaut and take him aboard, after which it flew back to the recovery ship with the capsule on the hook.

Immediately after the astronauts selected for MR-3 began their training, the press began speculating which of them would make the flight. The media favorite was the charismatic John Glenn. To the popular press there was no doubt that he would carry out the mission. The more intellectually inclined media believed that it would be Virgil Grissom. The reason was obvious: Grissom came from the air force, and the Department of Defense had just transferred responsibility for all future military space missions to that branch of the service. After taking into account inter-service rivalry, it also became plausible for the other correspondents to view Grissom as the most likely candidate.

On Tuesday, May 2, everything was ready for launch. Only the weather refused to cooperate. Wearing his space suit, Alan Shepard waited in Hangar S for three hours to be driven to the launch pad. Finally it had to be called off. Resumption of the countdown was put off until Thursday. Only then did Robert Gilruth inform the American public of the true identity of the first American astronaut.

The weather was no better on Thursday, but the outlook for Friday was better. The countdown was resumed at 2030 on the evening of May 4. Near midnight there was an unplanned interruption for the installation of the pyrotechnics, servicing of the hydrogen-peroxide system and to give the launch team a break. The countdown was resumed in the early morning hours of May 5.

The clock stopped two and a half hours before the scheduled launch time of 0700 in the morning. And that is how it was planned. They wanted to ensure that the checkout of the space vehicle was completed before they began bringing the astronaut to the pad.

Shepard was wakened at 0110. He showered and shaved and put on a bathrobe. Then he had breakfast with his physician William Douglas, his colleague John Glenn and several members of the operations team. He had been on a low-residue diet for three days, and on this morning breakfast consisted of orange juice, a filet mignon wrapped in bacon and scrambled eggs. At 0240 he underwent a final medical examination. Then the biosensors were stuck to his body, in places that had been marked with tattoos weeks earlier. Then Joe Schmitt, the suit technician, came and helped him put on his space suit.

Shepard boarded the transfer bus at 0355. On the way to the launch pad he lay on a reclining seat, while Joe Schmitt ventilated his suit with oxygen. When the bus had reached the launch pad, Schmitt put on the astronaut's gloves. As this was happening, Gordon Cooper gave him a report on the status of launch preparations.

At 0515, Shepard climbed the stairs to the lift gantry, in his right hand the portable climate-control system that he needed so as not to become overheated in the closed spacesuit. Five minutes later he climbed into the spacecraft. Schmitt tightened the straps. Then he solemnly shook Shepard's gloved hand. "Happy landings, Commander!" the gantry crew shouted in chorus. For Alan Shepard this was the most dramatic moment of his thirty-seven years, a moment he would remember with great clarity for the rest of his life.

The countdown was resumed. He began breathing pure oxygen at 0625, to free his blood of nitrogen and prevent a possible embolism. Fifteen minutes before launch the sky closed in and the flight director ordered a hold to wait for better photographic conditions.

Shepard passed the time by looking through his periscope. While he waited for visibility to improve, there was another hold and a 115-Volt rectifier in the launch vehicle's electrical system had to be replaced. This interruption lasted fifty-two minutes. Then the countdown was set back to the thirty-five minute mark. When it reached the fifteen-minute mark there was a malfunction in the IBM 7090 computer at the Goddard Center in Maryland. This required a complete reboot, a protracted procedure with the large computers of the day. With all these interruptions, an additional two hours and thirty-four minutes were added to the normal duration of the countdown. From there the countdown progressed without interruption to T 0.

Two minutes before launch, radio communications between the astronaut and ground control in the blockhouse next to the pad switched to Deke Slayton in the Mercury Control Center. Overhead Wally Schirra circled Cape Canaveral in his F-106, waiting to follow the Redstone and Shepard as high as he could. By the time the second to launch arrived, Shepard had been in the capsule four hours and fourteen minutes.

Shepard later admitted that, because of his excitement, he had noticed little of the final countdown except for the launch command. In one minute his pulse rate rose from eighty beats per minute to 126. At least as excited were the operations team, the press corps on the Cape and the millions of people watching the launch on their television sets.

Just prior to entering the space capsule, Shepard says goodbye to Virgil Grissom. In the center, to the left of Shepard's helmet with the white cap, is John Glenn.

THE FLIGHT OF FREEDOM 7

From the corner of his eye, Shepard saw the gantry tip away. He raised his hand to start the mission timer and was surprised by how gentle the liftoff felt and the clarity with which he heard Deke Slayton's voice from the control center. All of his reports were acknowledged without return questions. The quiet flight lasted forty-five seconds, then the vibrations began, at first soft then ever harder.

Shepard was prepared for this. He knew that the space vehicle was in the transonic range. The shaking became rougher and harder the closer he came to the maximum dynamic pressure. Eighty-eight seconds after liftoff the vibrations were so strong that Shepard's helmet was rattling so hard against the contour couch that he could no longer read the instruments. The noise level was extremely high, higher than he had expected, but bearable. Shortly after passing through the zone of maximum dynamic pressure, however, the noise and vibration disappeared and the flight became much smoother.

Cabin pressure inside Freedom 7 was held at 380 millibars, exactly as preprogrammed. Shepard calmly passed his reports, even when pressure rose to more than 6 g at two minutes after launch. Two minutes and twenty-two seconds after leaving the launch pad the engine shut down. The astronaut's speed was now 5,157 miles per hour. The flight path was one degree off nominal, which meant a deviation from the maximum flying altitude of less than 1.25 miles. After burnout, Shepard heard the escape tower fire and looked out the hatch, hoping to see the smoke trail.

But he couldn't see anything. The green tower jettison lamp on his instrument panel was the only indication that the pylon was gone. Shepard had been subjected to a maximum pressure loading of 6.3 g during the ascent. On the outside of the capsule the temperature had risen to 221 degrees Fahrenheit, while inside it was ninety-two. Inside his pressure suit the temperature was seventy-six degrees.

Two minutes and thirty-two seconds after launch, Shepard reported to Slayton that the capsule had separated from the booster. Three minutes after liftoff, the automatic attitude control system oriented the capsule with the heat shield in the direction of flight. There was a brief pendulum motion during the maneuver, but it stopped when the attitude control system automatically cut in. Now, almost at the apogee of his parabolic flight path, Shepard turned to his most important task: to determine whether he could manually control his vehicle's attitude in space.

He switched the attitude control system to manual. Individually, axis by axis, he tested the movements. First the pitch angle, which he was able to control by moving the joystick forward or backward. His first action was to position the space vehicle in precisely the right position to fire the retro rockets, at an angle of thirty-four degrees to the local horizon. While Shepard controlled the pitch angle, the control automatically system regulated the yaw (left and right) and roll angles of the spacecraft. When Shepard ultimately took control of all three axes, he was pleasantly surprised that the spacecraft's behavior was very similar to what he had learned in the procedures trainer.

Then he tried to observe the scenery below him. He noticed that the grey filter had been inserted into the periscope. That was due to his own mistake. During the hours of waiting on the launch pad, he had installed the filter to shield his eyes from the blazing sun. He had meant to remove the filter as soon as he retracted the periscope but had then forgotten.

He reached for the rotary switch to remove the filter, but in doing so the pressure gauge, worn on his left wrist like a wristwatch, struck the flight abort lever hard. Startled, he cautiously withdrew his hand. Until the end of the mission therefore, he saw everything—at least through his periscope—in black and white.

The launch of Freedom 7.

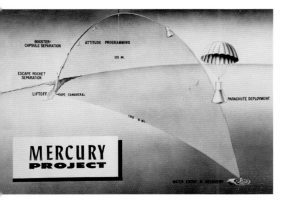

This contemporary illustration shows the flight path of Freedom 7.

Soon after Shepard had passed the apogee of the parabolic flight path at an altitude of 115 miles, it was envisaged that he would switch to fly-by-wire mode. He used the manual controls to change the capsule's attitude, employing the automatic system's hydrogen-peroxide jets. Shepard was supposed to manually position Freedom 7 for ignition of the retros, and as he did so he noticed that the pitch angle was too low—only about twenty degrees instead of the required thirty-four degrees. He corrected the pitch angle, even before he changed the yaw and roll positions. Then he went back to automatic mode and fired the three retro-rockets one after the other.

Immediately afterwards pieces of debris and retaining straps flew past the porthole, suggesting that the retro pack had been jettisoned. Shepard looked at his instrument panel but saw no confirmation. Seconds later, however, Slayton advised him that telemetry had reported jettisoning of the retro pack. As a precaution, Shepard also pushed the button to manually jettison the pack, and only then did the light come on.

Now on the descending arm of the parabolic flight path, and falling back to earth, Shepard reengaged fly-by-wire mode. He noticed a tendency to over-control the spacecraft. Then he switched back to the automatic system. Immediately afterwards

the periscope was retracted automatically as Freedom 7 approached the denser atmospheric layers.

On the way down Shepard tried to recognize stars through the two portholes but was unable to do so. Soon afterwards the indication appeared that showed the buildup of pressure. These deceleration values rose quickly and ultimately reached a maximum of 11.6 g. At the same time the spacecraft began to oscillate. Shepard reengaged the automatic control system and concentrated on the descent procedures.

Actually he was supposed to read out indications from the altimeter between seventeen and fifteen miles, but Shepard forgot because his rate of descent was greater than expected, the pendulum motions of the capsule were considerable, and he thought about jettisoning the drogue chute. In fact this occurred relatively late, at a height of four miles, and resulted in the pendulum movements stopping immediately.

After the descent had stabilized, he was able to extend the periscope again and he saw the drogue chute above him. The outside air valve opened at three miles. From then on, the internal and external pressures were adapted to match one another. The antenna canister was jettisoned at three miles, which triggered the release of the main parachute. Shepard could clearly see that the parachute was initially reefed and took several seconds to deploy its entire diameter of just under eight inches. The critical moment had passed and Shepard could begin preparing for the recovery.

As he descended toward the water at a rate of thirty-three feet per second, he pressed a switch to pump the remaining hydrogen peroxide overboard. A look at the instrument console showed a green light that signaled that the pneumatic landing bag had extended. Then, for the control center on the Cape, he disappeared beyond the radio horizon.

The astronaut used his remaining time to release his knee straps, open his helmet visor and separate his suit's hose fittings from the spacecraft. Then he struck the water, a rather harsh impact it seemed to him. The capsule listed about sixty degrees to the right—from the astronaut's point of view. Then he released the parachute. Water lapped over the portholes, and Shepard could see the fluorescent dye spreading quickly. He checked to make sure that the landing hadn't caused any leaks. He found none; the capsule was dry. Freedom 7 now also began slowly moving into an upright position.

The helicopter of Marine Air Force Group 26, stationed on the aircraft carrier *Lake Champlain*, was already waiting in the air. The primary helicopter flown by pilot Wayne Koons and copilot George Cox had observed the capsule during the last five minutes of its descent to the water's surface. Immediately after splashdown, Koons moved his helicopter into position to pick up the capsule and pilot. As he connected the cable to the recovery loop, looking at Freedom 7 Cox noticed that the high-frequency antenna was not in its proper position.

Koons lifted the capsule partway out of the water and waited for the pilot to climb out. Suddenly the high-frequency antenna snapped up, struck the helicopter, dented the outer skin and then broke off. There was no other damage. Shepard advised Koons that he would climb out as soon as Freedom 7's hatch was above the waterline.

While Shepard moved into a sitting position, Koons asked again if he was ready. He reported that there was still water lapping close to the hatch. The helicopter raised the capsule a little farther from the water and Shepard unlocked the hatch. Then he crawled over the sill, reached for the horse collar and placed it around him. He then gave the pilot the sign. With the capsule on the hook and Shepard on board, the pilot flew his helicopter back to the *Lake Champlain*.

When Shepard stepped onto the deck of the carrier, it had been just eleven minutes since he had touched down on the water and just thirty minutes after he had left Launch Complex 5 in Cape Canaveral.

On board the *Lake Champlain*, Shepard was first given a medical examination. Many doctors had feared that even a few minutes of weightlessness would cause disorientation and mental confusion. None of this had happened.

Shepard had come through the flight in excellent condition and not only that: he had also proved that a man could work sensibly and purposefully in the conditions encountered during a space flight. From beginning to end, the flight had been a complete success. The impetus effect of this mission on the American space program was immense.

On May 8, Shepard received the Distinguished Service Medal from President Kennedy at the White House. On May 25, 1961, President Kennedy gave his famous speech to both houses of the US Congress in which he spoke of the future of US space travel and its position with respect to its rival, the Soviet Union. The last four sentences of his speech read: "… we cannot guarantee that we shall one day be first, we can guarantee that any failure to make this effort will make us last. We take an additional risk by making it in full view of the world, but as shown by the feat of Astronaut Shepard, this very risk enhances our stature when we are successful. I believe this nation should commit itself to the goal, before this decade is out, of landing a man on the moon and returning him safely to the earth. No single space project in this period will be more impressive to mankind, or more important for the long-range exploration of space; and none will be so difficult or expensive to accomplish."

Mission Data	
Mission Name	MR-3
Callsign	Freedom 7
Date	May 5, 1961
Launch Site	Cape Canaveral, Launch Complex 5
Launch Vehicle	Redstone
Spacecraft	Mercury Number 7
Crew	Alan B. Shepard
Spacecraft Weight	2,844 lbs
Flight Path	Suborbital
Maximum velocity	1,783 mph
Flight duration	15 minutes, 22 seconds
Flight distance	303 miles
Flight path apex	117 miles
Landing site	western Atlantic
Recovery ship	USS Lake Champlain

THE SINKING OF THE LIBERTY BELL

After Alan Shepard's historic flight, Virgil Grissom and his backup John Glenn quickly returned to their own preparations. Production capsule 11 had been planned for use in the MR-4 mission. It had left the McDonnell plant in May 1960 and was the first Mercury capsule with a large window directly in the pilot's sightline. The spacecraft was also equipped with a number of other changes compared to Shepard's vehicle. The most important of these was the explosively actuated hatch. This piece of equipment was added in response to a suggestion by the astronauts. Originally, the only exit procedure required the pilot to crawl through the narrow cylindrical antenna and parachute compartment at the tip of the capsule. This was a complicated maneuver that required the removal of a pressure bulkhead. All of the astronauts had found it very difficult to work their way through this extremely narrow tunnel to the outside. Recovery of an injured astronaut would have been impossible. To open the hatch from the outside, several metal shingles of the outer skin had to be removed and then seventy bolts loosened.

The McDonnell engineers went back to work and ultimately came up with two types of exit hatch. One was equipped with a locking mechanism. This version was used for Ham's MR-2 flight and Shepard's MR-3 mission. The other model was the explosively actuated hatch. The mechanically complicated locking version had one decisive disadvantage: it weighed more than sixty-six pounds. That was too much for the orbital version. The design of the explosively actuated hatch was exactly the same as the original model. It had the same seventy bolts, each with a diameter of one-quarter of an inch. A 0.05-inch hole was bored into each of the screws to create an artificial weak spot. Then a recess was made around each screw, into which a small explosive charge was placed. When they were detonated, the bolts sheared off and the hatch popped out.

There were two ways of activating the explosively actuated hatch during recovery. About eight inches from the astronaut's right arm there was a pushbutton with a thickened end. The pilot had to remove a safety pin and then exert approximately six pounds of force to depress this button. That set off the explosive charge and blew the hatch a distance of about twenty-six feet. It was possible to activate the button with the pin still in place; however, this required about forty-four pounds of force. The recovery personnel could blow the hatch from the outside by removing a small panel from the capsule and pulling a lanyard. The explosively actuated hatch system weighed a total of just twenty-three pounds.

The new trapezoid-shaped window was warmly welcomed by the astronauts. It replaced the two ten-inch diameter portholes, which were unfavorably positioned. The window was now directly in the pilot's sight line. It gave him a field of view of thirty degrees in the horizontal and thirty-three degrees in the vertical.

The manual control had also been improved. It was now equipped with a so-called rate stabilization control system. This greatly improved handling of the space vehicle, for the astronaut could determine the magnitude of the attitude control movements himself by turning the control. Literally speaking, he was able to "give gas" or "let up on the gas" instead of maneuvering into a certain position with linear, equally strong impulses and neutralizing the kinetic energy that was built up with counter-impulses once in the target position.

These and a series of other changes took several weeks more than planned. In mid-July 1961, Grissom announced that his space capsule would be called Liberty Bell 7. The name of the freedom bell was known to every American and it was almost a synonym for freedom, and the number 7 had become naturalized as the "Mercury number" that he retained it.

There were also several modifications to the space suit, in particular with respect to mobility. A urine collector was new. Shepard, who spent a total of more than nine hours in his suit, had been forced to relieve himself in the suit when he was forced to wait hours for the launch. Grissom did not find the new installation particularly comfortable, but it did help resolve an annoying condition that had not received sufficient attention before.

Shepard had been overloaded with tasks during his five minutes of weightlessness. Grissom was supposed to get the opportunity to learn something about the visual possibilities of the new window, especially whether it was possible to recognize landmarks in order to use them as navigational reference points. As far as laying down the flight path went, the mission was a repeat of Shepard's flight.

Redstone booster number 8 arrived at Cape Canaveral on June 8, and immediately afterwards the task of mating it and the capsule began. Rocket and capsule were declared operational on July 15 and the recovery fleet moved into position. The launch was scheduled for July 18, but on the 16th this was pushed back to July 19, because the recovery area was forecast to have a low overcast on that day.

Early on July 19, Grissom was awakened by Bill Douglas and the launch day routine began, even though the weather at the Cape did not look particularly good. At five o'clock Grissom was at the gantry and climbed into the space vehicle. The countdown proceeded without interruption until ten minutes and thirty seconds before launch. Then a waiting period was begun, to wait for a hole in the overcast.

The wait proved to be in vain and the launch was cancelled and postponed for forty-eight hours.

On July 21, the weather was again less than ideal but still better than two days earlier, and so launch preparations began.

The countdown proceeded smoothly until during checkout a technician noticed that one of the seventy hatch bolts could not be screwed in straight. The

Alan Shepard is lifted onto the helicopter. Copilot George Cox's hand is still visible, and beneath Shepard, who is in the "horse collar," is Freedom 7.

The same scene photographed from another helicopter.

countdown was delayed for thirty minutes. The launch crew finally decided that the remaining sixty-nine bolts provided enough security. The countdown was resumed, but later it was interrupted twice, for a total period of one hour. Finally, at 07:20, the launch took place.

Like Shepard, Grissom was also amazed by the initial smooth ascent. But after about thirty seconds the vibrations began and quickly increased in intensity. Through the use of additional insulation in the attachment section between the capsule and launch vehicle, this time the vibration was damped enough for Grissom to be able to read his instruments. After about a minute, observers on the ground lost sight of the rocket as it entered a thin layer of cloud.

Two minutes and twenty-two seconds after leaving the launch pad, launch vehicle cutoff took place. Grissom's attention was now occupied by the release of the escape tower. In this phase between very high-pressure loads and weightlessness, he had a strong feeling of tumbling and for a moment lost orientation.

The capsule remained attached to the Redstone for ten seconds after burnout, then the Liberty Bell 7's posigrade rockets fired and the capsule was free. Then followed the automatic turn maneuver with the heat shield in the direction of flight. Grissom found it difficult to concentrate on his tasks as he constantly had the urge to look out the window.

He reluctantly turned to his instruments and his tasks with the controls. He found the capsule's reactions to manual inputs very sluggish compared to the reactions he was used to from the procedures trainer. Then he turned on the new rate command control system and found the response characteristics perfect. He did notice, however, that fuel consumption with this system was very high.

When Liberty Bell 7 passed the apex of the parabolic flight path at an altitude of 118 miles, it was time to place the space vehicle in the reentry position. Firing the retro-rockets gave him the distinct feeling that he had changed his direction of movement and was now flying forwards instead of backwards. Reentry posed no problems for Grissom. He was aware of the oscillations after pressure built up, but only from his instruments. In the meantime he reported his instrument readings to mission control.

The drogue chute was deployed at an altitude of five miles. The main parachute was released at three miles, about 985 feet above nominal height. Grissom saw the main chute deploy and noticed an L-shaped tear in the canopy about six-inches long

The window in the Mercury capsule is clearly visible here. Prior to this there were only two small portholes. Here John Glenn assists as Virgil Grissom climbs aboard.

and a two-inch hole. This worried him slightly, but the holes did not grow in size. Then he pumped his remaining hydrogen peroxide overboard and continued giving mission control readouts from his instruments.

A dull thump confirmed that the pneumatic landing bag had deployed in preparation for contact with the water. Grissom disconnected his space suit's oxygen hose from the supply system and opened the suit, but left the ventilation hose attached. The impact was gentle, as expected, but the capsule had such a list that Grissom was lying with his head facing almost down. Slowly, however, it began righting itself. When the window was above water, Grissom released the reserve parachute and pressed the button to activate the recovery system. Liberty Bell 7 rolled heavily in the waves, but appeared to be watertight.

The astronaut removed his helmet and prepared to climb out. He was unable to completely roll out the neck dam. He fumbled with the collar for a while

Grissom's Redstone launch vehicle is readied for launch.

to ensure that the suit would not take on any water, in case he should have to get out of the capsule quickly. At that time, the helicopter was already on its way. It had taken off when Liberty Bell 7 launched and had seen the capsule drifting beneath its parachute.

Now it was still two miles from the splashdown point. Lieutenant James Lewis, pilot of the primary recovery helicopter with the call sign Hunt Club 1, radioed Grissom to ask if he was ready to be picked up. He asked that the helicopter wait five minutes to allow him to complete making notes on his final instrument readings. That was a little difficult with the grease pencil and his bulky gloves. As well, the suit ventilation, which was still connected to the

capsule, was blowing on his neck constantly. Grissom solved that problem by sticking one finger between his neck and the rubber collar to let the air out.

When he had finished his notes, he removed the pin from the explosively actuated hatch and asked John Reinhardt, the helicopter's copilot, to initiate the recovery. As Reinhardt was in the process of cutting off the antenna in order to get to the recovery loop, he saw the hatch fly off, skip across the waves and finally sink. The next thing he saw was Grissom, who climbed out of the capsule and in his space suit swam a few yards away.

Instead of now turning their attention to Grissom, the crew of Hunt Club 1 continued to concentrate on removing the antenna from the space capsule and getting the Mercury on the hook. This prioritization was an ingrained reflex from recovery exercises with the astronauts. The training had taken place on the beach in Virginia, and the helicopter pilots had seen that the astronauts enjoyed swimming in the water.

By then the capsule had filled with water and continued to sink while Reinhard was busy pulling the hook through the loop. To stay at the height of the hook the helicopter had to go ever lower, until finally all three wheels were in the water. Liberty Bell 7 was now completely under water and out of sight, but when the helicopter climbed slightly the cable tightened and it was obvious that the capsule was hanging on the hook.

Reinhard then prepared to lower the horse collar to the astronaut. At that second an instrument on Lewis' instrument panel showed that the engine was overheating. Lewis considered the effects of an imminent engine failure, instructed Reinhardt to retract the horse collar and called in a second helicopter to take over Grissom's recovery.

Meanwhile, the astronaut noticed that the helicopter was having difficulties raising the almost sunken spacecraft. He swam back to the capsule to see if he could help, but he found the cable correctly attached. Then he looked up to grab the horse collar and saw that the helicopter was turning away.

Grissom now noticed that he was not as high in the water as before. He realized that air had been escaping through the neck dam the whole time. The more air he lost, the less buoyancy he had. Unfortunately, on top of everything he had forgotten to close his suit's air intake valve. The suit quickly filled with water and the waves began closing over his head more frequently. It would be even worse when the second helicopter came and he was pushed under water by downwash from the rotors.

Grissom now had to struggle with fear. He was furious, because obviously no one was worrying about him and there were no rescue divers to be seen anywhere. Then he saw a familiar face in the second helicopter: it was Lieutenant George Cox, who had been present for the recovery of Ham the

chimpanzee and Alan Shepard. Although he was barely able to keep himself above water, the sight of Cox gave him new confidence.

Cox recognized what was happening and threw the horse collar right at Grissom, who somehow wriggled into the loop and at the end hung backwards in it. But now was not the time to look good. Before the line tightened, Grissom sank below the surface for a few seconds, and it was no wonder that immediately after he was pulled aboard the helicopter he reached for the nearest life vest and wrapped his fingers around it. Grissom had only been in the water for about four and a half minutes, but—as he later said—"it seemed like an eternity to me."

Reinhard's helicopter, meanwhile, was still struggling with Grissom's water-filled capsule and had raised it partly out of the water. All of the water did not all run out of the pneumatic landing bag, and so the capsule stayed in the water as if it had dropped an anchor. The weight indicator in the helicopter confirmed that a weight of almost 2.5 tons was hanging beneath the machine, 1,100 pounds over the maximum load. Reinhard saw his blinking red warning light and decided that it made no sense to lose two flying vehicles. And so he unhooked the capsule.

Grissom underwent a medical examination on the deck of the carrier. He was exhausted but insisted on the debriefing. Later he said that he was very grateful to Wally Schirra for coming up with the idea of the neck dam, for it had probably saved his life. As a result of this incident, it was decided to conduct more recovery training in the future. In Grissom's case, the last recovery exercise was four months back.

For the most part he found the capsule functions within acceptable limits. He did however criticize a number of points. There were far too many belts to connect and tighten, the instrument lighting was too weak, the oxygen consumption rate too high, the urination system impractical and the rate control system's fuel consumption rate excessive.

There remained the problem of the explosively actuated hatch. It could never be clarified.

The launch of Liberty Bell 7.

Before the mission, the system had been extensively tested under the toughest conditions by its maker, Minneapolis-Honeywell. After the mysterious incident, which with a little more bad luck could have cost Grissom his life, the hatch was again put through its paces. It never failed. Suspicion therefore turned towards Grissom himself and the possibility that he had unintentionally activated the hatch.

There was much conjecture, from flashover voltage between the helicopters or the capsule to the possibility that the pull cord was activated from outside, possibly by the pneumatic landing bag, which banged against the capsule in the fairly rough seas. In the end, however, one factor supported Grissom. He was the only astronaut who had not injured his hand operating the explosively actuated hatch. Glenn, Schirra and Cooper each sustained bruising and minor cuts as a result of operating the explosively actuated hatch. From then on, the astronauts were not supposed to touch the safety pin until the helicopter had the capsule on the hook and the line to the helicopter was taut. The procedure was unnecessary, however. Liberty Bell 7 was the last manned Mercury flight in which the capsule was pulled from the water by helicopter.

Mission Data	
Mission Name	MR-4
Callsign	Liberty Bell 7
Date	July 21, 1961
Launch Site	Cape Canaveral, Launch Complex 5
Launch Vehicle	Redstone
Spacecraft	Mercury Number 11
Crew	Virgil I. Grissom
Spacecraft Weight	2,844 lbs
Flight Path	Suborbital
Maximum velocity	5,169 mph
Flight duration	15 minutes, 27 seconds
Flight distance	302 miles
Flight path apex	118 miles
Landing site	western Atlantic
Recovery ship	USS Lake Champlain

FROM THE TITOV SHOCK TO GUEST PERFORMANCE OF THE SCOUT

Even before Grissom's flight, the Manned Space Center considered not going through with mission MR-5, but it wanted to evaluate the data gained from MR-4 before making a decision. If they could accelerate the Mercury-Atlas, so the thought went, then that would move America forward in the race with the Soviets. After all, an astronaut circling the earth three times would trump a cosmonaut with just one earth orbit.

On August 7, 1961, they could safely disregard this hope. That was the day when, at nine o'clock Moscow time, twenty-six-year-old Soviet Lieutenant German Titov blasted into orbit aboard Vostok 2. He remained aloft until ten the following day and then returned to earth.

The Titov flight finally put to rest the plan to still fly mission MR-5. And after the date from the Grissom flight had been reviewed, there was a general consensus that not much would be gained by another suborbital flight. So on August 18, NASA announced that the Mercury-Redstone program had achieved its objectives and that all further suborbital missions had been cancelled.

Completely soaked and exhausted, Grissom on board the USS Randolph.

The USSR was not forthcoming with information about the flight of Vostok 2, but some managed to filter through. One point in particular gave the people at the Manned Space Center cause for concern. Titov had obviously had to battle symptoms that he himself characterized as "very similar to seasickness." His was thus the first report of a condition that would later be called space sickness. This "illness" began with symptoms like nausea and vomiting, and the critical point was that a space traveler forced to vomit in a fully closed spacesuit faced the danger of aspirating.

The Soviet Union now had two orbital flights on its account. NASA still had none. And the news that had to be given to the impatient American citizenry was not rosy. The day that Titov returned from orbit, NASA had to announce that the first American manned orbital mission would slide to January 1962.

The reason for that was that flight qualifications were still outstanding. The first was the initial orbital test of the Mercury-Atlas combination with mission MA-4. Once again, the astronaut on board was to mimic the crewman simulator, which had been on the failed mission MA-3. That would be followed by an orbital mission, during which the track following and communications network would be tested. This would be the only flight in the program that was to be carried out with a Scout rocket. And finally there would be the orbital test with a chimpanzee on board, which was to fly the MA-5 mission. According to the plan, only after this could they risk a manned flight.

Number one on this list was mission Mercury-Atlas 4, the fifth flight test by an Atlas with a Mercury and a repetition of the failed mission MA-3 of 25 April. Atlas number 88D was envisaged as the launch vehicle, but it failed its acceptance inspection on June 30 and had to be sent back to the factory. It did not arrive at the Cape until July 15. Mercury production unit number 9 was originally envisaged as the capsule, but after capsule number 8, whose launch vehicle had been blown up over the Cape on April 25, was found to be in good condition, it was sent back to St. Louis for overhaul. The idea was not such a good one, for the measures proved more expensive than hoped and lasted into August. The overhauled capsule was given the number 8A. Planned launch date was August 22.

On the day of the flight, the Air Force Space Systems Division in Cape Canaveral telephoned and reported that soldering residue had been found in transistors on the Atlas production line at Convair. As it turned out, this type of transistor was used not just in the launch vehicle, but also in the Mercury space vehicle itself. There was no choice but to delay the launch and replace all suspect components. The entire process lasted until September 1.

MA-4 was supposed to be placed into an orbit with a perigee of ninety-six miles, an apogee of 155 miles and an inclination to the equator of 32.5 degrees. The capsule was plastered with numerous sensors inside and out. Microphones at the astronaut's head level would pick up the noise level. The vibration level would be measured by seven different sensors, most of them in the adapter region between launch vehicle and capsule. Four standard radiation detectors would be taken along and placed in various parts of the capsule. There were also two broadband measuring devices that were capable of observing the entire radiation spectrum as well as the depth of penetration of the radiation. Flight data was transmitted to earth over two separate, redundant channels. The sensor data was saved in the capsule using magnetic tape recorders.

The flight was also extensively documented photographically. On the left side of the cabin there was a camera that was supposed to take approximately 20,000 photographs until splashdown. An earth observation camera was to take about 600 photos and 10,000 more were to be taken by another camera through the periscope.

On launch day, September 13, 1961, launch conditions were ideal. The countdown was halted for half an hour ninety minutes before launch to

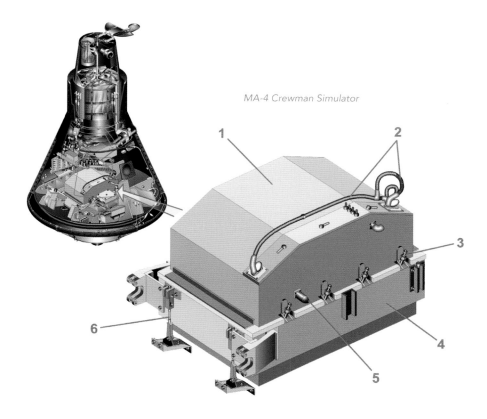

MA-4 Crewman Simulator

1

2

3

4

5

6

The crewman simulator was a machine that imitated some of the metabolic functions of an astronaut in order to test the Mercury capsule's life support system in flight.

1. Crewman Simulator. 2. Life support system attachments. 3. Screw connections to instrument pallet. 4. Instrument pallet. 5. Electrical connection. 6. Cabling.

replace a broken screw in one of the capsule's exterior panels. Fueling with liquid oxygen began at 08:30. At 08:57 there was a problem with data transfer from the tracking station in Bermuda and the countdown had to be reset to three minutes and thirty seconds before launch. At 09:04 the time came. The vernier rockets ignited, followed by the main engine, and MA-4 was on its way into orbit.

There was considerable vibration during the first twenty seconds of flight, but Atlas 88D was the first of a series of rockets with a reinforced structure, so this load caused no problems. Fifty-two seconds after liftoff a rectifier failed, however, the backup unit immediately took over. Flight monitoring showed that the flight path was initially 0.75 degrees too high, but towards the end of the powered phase 0.14 degrees too low. Booster burnout took place 2.5 seconds too early; nevertheless velocity at that time was 56 mph too high. This was compensated for by an early shutdown of the sustainer.

All of these deviations played out within normal operating parameters and presented no problem.

Perigee was one mile and apogee eleven miles below planned values, but these values too were well within limits. And so, immediately after burnout, the computer in the Goddard Center gave its "go" for continuation of the mission.

Following separation from the booster, the maneuver that was to place the heat shield in the direction of flight began. A problem arose during this maneuver, because the space vehicle made several pendulum movements about all three axes before it finally ended up in the desired position. Instead of twenty seconds, the positioning maneuver took fifty seconds, and instead of two pounds of fuel, ten pounds were used. An open electrical connection in the gyroscope was later discovered. It had been responsible for the pitch rate. The problem was not so severe as to pose a threat to the mission.

The next problem was the high oxygen consumption. Virgil Grissom had previously reported this and it occurred again. This was combined with a tendency toward high cabin pressure. Instead of 380 millibars the figure was 415 millibars. Near the end of the mission the primary oxygen supply was used up and the system had to be switched to reserve. The crewman simulator functioned well throughout the entire mission. It consumed oxygen— at the same rate as a man would—and gave off carbon dioxide, heat and moisture.

There were several minor problems with the attitude control when, towards the end of the mission, two of the steering jets failed, one for yaw angle control and one for roll rate. The redundancies on board, however, were sufficient to compensate for the loss of these components. Communications between the capsule and the ground stations were satisfactory. There were only a few minor problems, which were attributable to lack of experience on the part of the operators.

One hour, twenty-eight minutes and fifty-nine seconds after leaving the launch pad, the first of three retrorockets fired near the Hawaiian Islands. As the capsule passed overhead, the tracking station at Guaymas, Mexico, and Cape Canaveral Control confirmed that it was in the correct position for reentry. The drogue chute opened over the Atlantic at an altitude of eight miles and the main parachute at two miles. The capsule splashed down at 10:55, 177 miles east of Bermuda. One hour and twenty-two minutes later it was picked up by the destroyer *Decatur*, which had been thirty-four miles from the landing site when the capsule splashed down. It took the capsule to Bermuda, from where it was flown to the Cape for examination.

The reason for the high oxygen consumption was quickly discovered. An astronaut on board could have corrected it manually. The vibration during liftoff had caused the power lever to slip from its detent, causing the fuel valve to snap open. The flow rate was so low, however, that the micro-switch that was supposed to initiate telemetry transmission did not function. Otherwise, there had been few problems. Several small cracks and dents caused by impact with the water, washers and screws floating through the cabin, which showed that the interior of the capsule was still not being cleaned well enough, an oily film on the window, and so on. MA-4 was judged an unreserved success.

Just five days after the success of MA-4, there was a significant organizational event for the new Manned Spaceflight Center. NASA administrator James Webb announced that the USA's manned spaceflight program was to move to a 1,000-acre facility in Houston, Texas. It was not until mid-1962, however, that the move was completed. During this busy period an episode in the Mercury test program occurred that was forgotten soon after its inglorious end.

The Mercury tracking network was completed in March 1961 and was awaiting comprehensive testing. This would, however, have required a satellite that simulated a Mercury spacecraft by following the same orbit as a capsule. A launch vehicle would also be required to place the satellite into orbit. This had not been foreseen in planning the program budget, however, and so it had to be the cheapest possible rocket and the most inexpensive satellite available. And so the NASA scientists looked to see

*Launch of Atlas D Number 88 with production
capsule 8A on the Mercury Atlas 4 mission.*

if the four-stage, solid-fuel Scout rocket, then in use by both NASA and the air force, could be used. The scout was called "the poor man's launch vehicle." It was composed of four solid-fuel rocket motors one on top of the other and had an all-up weight of just seventeen tons.

Abe Silverstein in NASA headquarters thought the test unnecessary, but the manager, who was responsible for all operational aspects of the program, thought the training mission a good idea. The affair would probably have soon been buried again had an inquiry to the air force not revealed that it had an extra Blue Scout (the military version of the Scout) without a mission. The air force showed itself to be ready not just to sell the rocket but to conduct the launch as well. And so the plan was authorized with the rather reluctant support of NASA central. The selected launch date was August 15, 1961.

To keep costs down, and because time left no other alternative, a separate satellite was not built. Instead a sort of angular extension was added to the fourth stage of the Scout. This container was used to house the communications equipment from Mercury capsule 14, which had flown on the Little Joe 5B test on April 28. The entire thing was powered by a battery with a capacity of 1,500 Watt hours. Designated MS-1, the box weighed 149 pounds.

The battery was capable of powering the electronics for eighteen or nineteen hours before it was exhausted. To extend its operating life, the equipment was to be powered down after three orbits. Then the data would be analyzed. Following the analysis everything would be powered up again for five hours. The procedure would be repeated towards the end of the flight. The goal of the exercise was to simulate three Mercury missions.

Once the mission had been authorized, work proceeded in typical NASA style: a concentrated effort rather than a dispersed one. That meant immediately acquiring a second Scout as a backup launch vehicle.

The mission profile was fixed at the beginning of July. MS-1 would lift off from launch complex 18-B, from which the Vanguard satellites had been

Mission Data	
Mission Name	MA-4
Date	September 13, 1961
Launch Site	Cape Canaveral, Launch Complex 14
Launch Vehicle	Atlas D (Number 88D)
Spacecraft	Mercury Number 8A
Crew	Crewman simulator
Spacecraft Weight	2,700 lbs
Flight Path	Orbital
Orbits	1
Perigee and apogee	97 x 154 miles
Flight duration	1 hour, 49 minutes, 20 seconds
Inclination	32.57 degrees
Landing site	western Atlantic
Recovery ship	USS Decatur

put into orbit. On July 25, the Scout was assembled and awaiting its payload. The maker had fallen behind, however, and it quickly became clear that a July launch would be impossible. The instrument package finally arrived at the Cape in early August, but then difficulties arose with the Scout's fourth stage. To save time, it was decided to use the fourth stage from the spare rocket. The conversion took more time and the reassembled Scout did not reach the launch pad until October 22. By then the MA-4 flight had long since taken place and had checked out the radio and telemetry systems. The Scout's mission had thus actually become redundant. But it was on the launch pad and so it was decided to go ahead with the launch on October 31 and give the people of the telemetry and tracking network a few additional training segments. The countdown went smoothly until the moment of ignition. Nothing happened, and the rocket sat motionless on the launch pad.

The ignition circuit was checked and repaired and on November 1 the Scout lifted off on the equally short and noteworthy mission MS-1. Immediately after liftoff, the rocket was seized by wild snaking movements and twenty-eight seconds after leaving the pad it broke up. The safety officer on the ground transmitted the self-destruct command, but it was actually unnecessary. The failure that had taken place was as simple as it was fatal. A technician had misconnected the wiring between the gyroscopes for pitch and yaw control, so that deviations in the yaw angle were fed to the pitch control and vice versa. Six months of planning had gone up in smoke in less than a minute.

Brief consideration was given to a second Mercury-Scout flight. Then the realization that the program had simply progressed beyond this mission gained the upper hand. Because of the fact that MA-4 had already flown and that MA-5 was imminent, the plan was dropped. Thus ended the equally brief and chaotic life of the Scout launch vehicle in the Mercury program.

Mission Data	
Mission Name	MS-1
Date	November 1, 1961
Launch site	Cape Canaveral, Launch Complex 18
Launch vehicle	Blue Scout II
Spacecraft	MS-1 satellite
Spacecraft weight	140 lbs
Flight duration	28 seconds

THE SCOUT

This economical, exclusively solid-fuel-powered launch vehicle was used by the American space program for three decades in a large number of variants. It was developed by NASA and built by Ling-Temco-Vought. The basic model was the Scout X1 and it made its first successful flight on October 10, 1960. It was a four-stage rocket with the following configuration:

- First stage: Aerojet General Algol
- Second stage: Thiokol XM33 Castor IA
- Third stage: Allegany Ballistics Lab
                   X-254 Antares
- Fourth stage: Allegany Ballistics Lab
                   X-248 Altair 1A

The Blue Scout II was a military version of the Scout X1. It was identical to the civilian basic version used by NASA with one exception. The fourth stage was concealed in a payload fairing with the same diameter as the third stage. The Blue Scout II was used a total of three times: twice for suborbital flights by the US Air Force and once in Project Mercury. While the suborbital flights were successful, the Project Mercury flight was a failure.

ENOS GOES INTO ORBIT

McDonnell production capsule number 9 incorporated all the modifications and changes that resulted from missions MR-3, MR-4 and MA-4. For this reason alone, the heads of the Manned Spaceflight Center wanted to insert another test flight into the program before the first manned orbital flight, MA-6.

MA-5 was to be carried out with Atlas D booster number 93. Convair had promised delivery for mid-August 1961, but then—already almost traditionally—there were a number of "reworks" and the date was postponed until October 9.

The chimpanzee Enos took over the job of the crewman simulator. He was also from the colony at Holloman Air Force Base, and his preparation had been similar to that experienced by Ham. Captain Jerry Fineg, the head veterinarian for the mission, described Enos as a "relaxed and cool type." The "replacement people" for Enos were, in this order, the chimpanzees Duane, Rocky and Ham. It was the same Ham who had flown on the Mercury-Redstone 2 flight.

Enos had four areas of responsibility to master. Problem one was the operation of two levers with which he was to switch off lights. If he forgot, or did it incorrectly, he received a mild electric shock, comparable to a gentle nudge. Twenty seconds after a green light lit up, at the earliest, Enos had to operate a lever in order to receive a drink of juice. Here there was no reminder if he did wrong. He would simply receive no juice. Problem 3 involved operating another lever. He had to press it fifty times to receive a banana chip. This exercise was also voluntary. If he didn't do it or pressed too many times, there was simply no banana chip. The fourth test required Enos to arrange symbols, discs, triangles and squares correctly. Here, too, there was a mild shock if he performed his task incorrectly.

It soon turned out that October 9 launch date could also not be met. Special capsule number 9 was proving to be fault-prone. A leak in the hydrogen-peroxide system forced a postponement, initially to November 7, then to the fourteenth and finally the twenty-ninth.

The countdown for MR-5 began on November 28, 1961. Eleven hours before launch, Enos was medically examined and prepared for liftoff. Five hours before launch he was placed in the capsule on his special pressure couch. Enos' condition was then monitored by telemetry and radio. He knew the procedure from many exercises and was relaxed. The only time he exhibited any agitation was at T-30 when the hatch had to be opened again because a switch in the capsule was in the wrong position. This halt lasted eighty-five minutes. The unplanned stop times ultimately cumulated to two hours and thirty-eight minutes and Walter Williams, the launch director, was visibly more impatient than Enos, who had already been enclosed in the capsule for seven and a half hours.

Launch took place at 10:08. The powered booster phase went well, although there was a series of minor anomalies. By the time the sustainer burned out, however, speed, angle of flight and altitude were almost perfect. The Atlas placed the space vehicle into an orbit with a perigee of ninety-nine miles and an apogee of 147 miles. This was slightly too low, but within limits. Separation of the spacecraft and the booster took place at precisely the planned time. This time the Mercury's turning maneuver took less than thirty seconds. Of the total of sixty-four pounds of attitude thruster fuel on board, the capsule used just six pounds for the turning maneuver. On MA-4 it had been ten pounds.

The chimp astronaut remained in good spirits. At the end of the propulsion phase he experienced acceleration forces of 7.6 g. As planned, he had begun his tasks two minutes prior to launch. During his flight he had to apply himself to a task twenty-nine times and made two errors. During task two, during which he had to wait twenty seconds for the

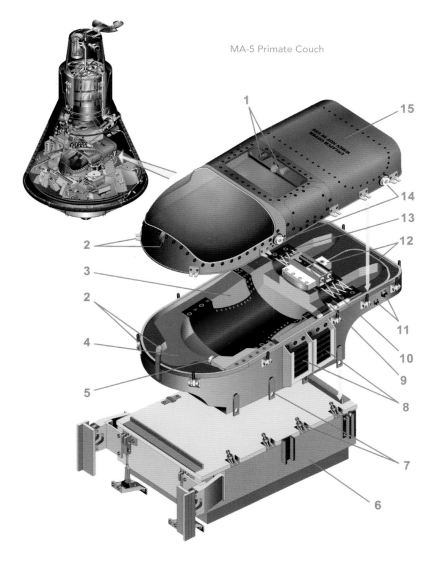

MA-5 Primate Couch

For their total of four non-human astronauts, the Project Mercury engineers designed the primate couch, a pressure compartment inside the pressure compartment that simulated a human astronaut's spacesuit. The experimental animals thus enjoyed the same level of safety as their human "colleagues."

1. Test equipment (switchboard). 2. Lock. 3. Seat depression. 4. Head rest. 5. Drinking system. 6. Instrument pallet. 7. Screw attachment to instrument pallet. 8. Filter. 9. Urine collector. 10. Blood pressure gauge. 11. Electrical connection. 12. Drinking water supply. 13. Leg restraint. 14. Life support system connection. 15. Primate couch cover.

Enos is prepared for launch.

capsule's space position to be correct. And so it went on, as the capsule was passed on from tracking station to tracking station as it circled the earth.

When MA-5 reached the vicinity of tracking station Canton Island, it was clear to the controllers that vehicle's attitude control system was allowing the vehicle to drift away from the correct position, only to correct it immediately afterwards. After the flight it was discovered that a metal chip had blocked the flow of fuel to one of the roll thrusters. This inactive thruster allowed the spacecraft to veer up to thirty degrees from the normal space attitude. The automatic stabilization and control system intervened each time and returned the capsule to zero degrees deviation. Then the game began anew. The capsule repeated this process nine times during the second orbit of the earth. Each of these maneuvers consumed half a kilogram of fuel.

Finally the cooling system in the environmental control system began acting up. Between the Canary Islands and Kano the temperature in the primate couch began to rise and the doctors on the ground slowly became worried. Enos' body temperature soon rose to thirty-seven degrees and finally to thirty-eight degrees, where it stabilized. The physicians decided that the mission could continue. Christopher Kraft, the flight director, was of a different opinion, however. It could be that the flight could continue from a medical point of view, but the erratic behavior of the attitude control system caused him concern. His engineers believed that with this condition, by the end of the third orbit there would not be sufficient fuel for a controlled reentry, which was very fuel intense even under normal conditions. He decided to continue the flight at least as far as California and informed Gordon Cooper at Point Arguello that this station might have to transmit the signal to the capsule to initiate the retro sequence. All of twelve seconds before the point at which a decision for a normal reentry after the second—instead of third—orbit was reached, Christopher Kraft decided to bring Enos back. And so Point Arguello transmitted the return command at the last second.

juice, his average waiting time was thirty-four seconds. He received his drink forty-seven times, about a half liter in total. Thirteen times Enos received banana pellets during task three. He had poor luck with exercise four. First of all he was not very good at this exercise—he managed just eighteen of twenty-eight symbols the first time around—and thus received ten electric shocks. During the second of four sessions the mechanism failed and he even received a shock if he pressed the correct lever. During the third and fourth sessions, therefore, he received thirty-six and forty-three shocks in succession. Despite the frustration he must have felt, he continued operating the lever.

Toward the end of the first orbit things began to get a little out of hand. First, ground control determined that for some reason the onboard clock was eighteen seconds fast. The old problem of the rectifiers overheating reappeared, the life support system gave contradictory information, and finally the capsule began making a series of curious direction changes.

The matter of the rectifiers at least caused no great concern. It had happened several times before, and it was known that it could be resolved by turning the system on and off. More serious was the matter of the curious movements of the capsule. While the tracking station in Muchea in Australia confirmed a deviation, another tracking station found the

With the exception of a repetition of the attitude deviation, which was caused by the inoperative thruster, the reentry went according to plan. In the primary splashdown zone were the destroyers *Stormes* and *Compton* and a P5M search aircraft. Three hours and thirteen minutes after launch and nine minutes prior to splashdown, the aircraft spotted the capsule swinging beneath its parachute at an altitude of just under 1.5 miles and informed both ships that they were about thirty-one miles away. The P5M circled over the capsule for an hour and fifteen minutes waiting for the *Stormes*. Then the destroyed arrived and pulled Enos and his spacecraft on board. Immediately afterwards the hatch was blown from the outside using the lanyard.

The capsule was examined closely on board the *Stormes*, later at the Cape, then at the Manned Spaceflight Center, and finally by McDonnell. All found that—apart from a few minor issues—it had survived the trip well. All of the problems that had arisen could have been corrected quickly if an astronaut had been on board. Enos had been weightless for 181 minutes and had mastered his tasks with aplomb. After being taken aboard the *Stormes* he ate two oranges and two apples. He had quite obviously survived his spaceflight in fine fettle. Enos retuned to the Cape on 1 December. There he underwent another medical examination and later was sent back to Holloman Air Force Base. Regrettably he died just nine months later, on November 4, 1962, of dysentery.

Enos had been under constant medical watch until two months before his death, when after care ceased. A pathological examination revealed that there was no connection between his death and the space flight.

"We're a Little Behind the Russians and A Little Ahead of the Americans"

A caricature from the year 1961.

Mission Data	
Mission Name	MA-5
Date	November 29, 1961
Launch Site	Cape Canaveral, Launch Complex 14
Launch Vehicle	Atlas D (Number 93D)
Spacecraft	Mercury Number 9
Crew	chimpanzee Enos
Spacecraft Weight	2,932 lbs
Flight Path	Orbital
Orbits	2
Perigee and apogee	99 x 147 miles
Flight duration	3 hours, 20 minutes, 59 seconds
Inclination	34 degrees
Landing site	western Atlantic
Recovery ship	USS Stormes

PREPARATION FOR JOHN GLENN

After the Enos mission, all of the flight qualifications were now completed. At the post flight conference for MA-5, therefore, the press representatives had just one question: who would be the first American in orbit? Robert Gilruth was ready for the question. He said that they had selected John Herschel Glenn to pilot the first American orbital flight. His alternate was Malcolm Scott Carpenter. The mission after that would be flown by Donald Kent Slayton. His alternate would be Walter Marty Schirra. Launch date was set for January 16, but the new year was just three days old when the media learned that this target date could not be met due to problems with the Atlas' fuel tanks and that the new date was January 23.

The capsule with the McDonnell serial number 13 was delivered to the Cape on August 27, 1961. Four months later, after an intensive checkout by the Manned Spaceflight Center team, on January 2, 1962, it was mated to Atlas 109D. Glenn's family gave the space vehicle the name Friendship 7.

The capsule and launch vehicle were thus almost ready for the historic mission. But what about the astronaut and his alternate? In the weeks and months leading up to the launch, their work plan was configured so that they would each reach their personal flight readiness status exactly on the launch date. It was a kind of personal countdown.

The training directive called for the astronauts to begin a comprehensive study of the on-board instrumentation eighty-one days before launch. This was important, because the Mercury capsules only resembled each other externally. There were modifications from mission to mission and each pilot had to get to know his capsule. At seventy-two days before launch, they each began spending at least three hours per day working in the procedures trainer and also the air-lubricated free-axis, or ALFA trainer in Langley. This was a motion simulator that had replaced the monstrous MASTIF.

In the procedures trainer the astronauts went through various mission profiles and many different emergency scenarios, beginning with an abort on the launch pad up to main battery failure immediately after orbit insertion. Such scenarios were only limited by the imagination of the training personnel who operated and programmed the simulator.

The month before the planned launch date marked the beginning of dress rehearsals. These saw the launch day played out, beginning with the waking of the astronauts, fitting them with biosensors, breakfast, putting on the space suit, transport to the launch pad and climbing into the space vehicle. This exercise was repeated one week later, but this time the gantry was rolled away. A third such exercise took place three days before launch, this time with a countdown that ran to launch point and in which Glenn's nominal five-hour mission was played out.

Glenn and Carpenter completed their training at the end of January. The continuing delays, however, forced them to continue their rushed training routine. Glenn spent a total of twenty-five hours in the space vehicle, during which time it was in Hangar S. Between December 13, 1961, and February 17, 1962, he spent almost sixty hours in the procedures trainer, far more than the required thirty. During that time he flew seventy simulated missions and was confronted with 189 system errors by his trainers. Glenn, Carpenter, Slayton and Schirra also took part in a recovery exercise that took place on the Back River at Langley.

The pilots were not the only ones who underwent intensive training. The recovery personnel also held a series of exercises, as did the personnel of the worldwide tracking and radio network. The tests reached such a scale that those responsible for the flight operation, Eugene Kranz and Christopher Kraft, feared that the performance of the participants would suffer if the flight did not take place soon.

NASA accredited about 400 journalists for the event. They eagerly took in the information prepared by NASA, conducted interviews with officials and hoped for additional bits of news and anecdotes with which to spice up their reports. The first launch party for all of these press people took place on January 27, a cloudy Saturday morning. January 23 had in fact been chosen as launch day, but day after day there had been further postponements on account of the persistent bad weather.

The countdown proceeded without problems until twenty minutes before launch, but to the observers who had a view of the sky it was clear: it was not going to happen on this day. Finally

mission director Walter Williams gave up. In this weather, camera observation of the launch would have been impossible. John Glenn had spent five hours in his spacecraft. The launch was postponed another four days.

Refueling of the Atlas 109D had just begun on January 30, when, during a routine opening of a drain screw, a mechanic discovered fuel in the space between the rocket structure bulkhead and insulation that separated the fuel and oxidizer tanks. This cost another ten days because the insulation had to be removed before the repair could be made, after which a new system checkout had to take place.

This delay, and it was not the first, caused difficulties for the recovery fleet. Twenty-four ships

The artist Cecilia Bibby painted the mission callsign chosen by Glenn's family on the side of his capsule.

Glenn and Carpenter during training for the Mercury-Atlas 6 mission.

and more than sixty aircraft, a total of 18,000 men around the world, had been at waiting stations for weeks and would now have to wait for an undetermined time longer.

Vice-Admiral John Chew, commander of the flotilla, therefore made his own calculations. He determined that, from the point of view of the recovery forces, the earliest possible launch date was February 13, and so he sent his people on leave.

Once again, however, the weather proved uncooperative. Not until February 19 did the sky finally pick up, as did the mood of the operations crew, and the countdown began. Glenn was awakened at 02:22 on February 20. He showered and then had breakfast: steak, scrambled eggs, orange juice and toast. At 03:05, he was given a quick medical examination by William Douglas.

At 04:27 Christopher Kraft, who was already sitting at the flight director's console, received news that the global tracking system was ready. Meanwhile, in Hangar S Douglas attached the bio-sensors to Glenn's skin, then suit technician Joe Schmitt began helping him into his suit. At 05:01 Mercury Control learned that Glenn was in the bus and on his way to the launch pad. He arrived at the pad at 05:17, twenty minutes behind schedule. The delay was inconsequential, because a problem had developed in the booster's flight control. The installation of a

reserve unit and its checkout took two hours and 15 minutes.

During the wait Glenn relaxed in the transporter. At 05:59 he left the van, saluted the onlookers and entered the lift that took him to the top of the rocket. At 06:03 he climbed into the spacecraft. Immediately after entry there was a problem with one of the bio-sensors, which had slipped. Installing it correctly it would have required opening the spacesuit again. After some consultation it was decided to leave it as it was, and the technicians began sealing the hatch.

At 07:10, by which time most of the bolts had been tightened, a broken bolt was discovered. Even though Grissom had flown MR-4 minus one bolt, it was decided this time not to launch. Walter Williams thus ordered the repair to go ahead. Removing all the bolts and bolting the hatch again took forty minutes. Glenn remained calm during this tense period. His pulse ranged from sixty to eighty beats per minute.

Installation of the hatch was finally completed and the cabin was pressurized. The leak test revealed a leak rate of thirty cubic inches per minute, which was well within acceptable tolerances. At 08:05 the countdown had reached T minus sixty minutes. Then there was another

Glenn climbs into Friendship 7.

Inside Friendship 7, John Glenn waits for launch.

halt of fifteen minutes to top the fuel tanks of the Atlas. Afterward, the countdown continued, but at T minus twenty-two minutes, by which time it was 08:58, a fuel ventilation valve stuck and another waiting period began.

The problem with the valve was fixed at 09:25. At 09:40 the countdown was stopped again, because the Bermuda tracking station reported a loss of power. It was several minutes before the system there was running again and the countdown could go into its final phase.

THE MISSION OF FRIENDSHIP 7

Launch took place at 09:47, three hours and forty-four minutes after John Glenn boarded Friendship 7. At the moment Atlas 109D lifted off, Glenn's heart rate rose to 110 beats per minute. A few seconds after liftoff the General Electric-Burroughs flight control system switched on the booster's radio transponder and steered the vehicle through the orbit entry window. About 100 seconds after leaving the launch pad, Glenn reported increasing vibration. After passing through the zone of maximum dynamic pressure the flight became smoother. Then, two minutes and fourteen seconds after liftoff, came the command BECO: booster engine cutoff. The Atlas' two external engines switched off and were jettisoned. Soon afterwards the escape tower was also jettisoned.

Towards the end of the sustainer's operation, vibration increased again and then stopped abruptly when it was shut down. The computer in Maryland calculated that the capsule had achieved a velocity that was just 1.25 miles per hour under the planned value. Inclination above the equator had been achieved to within one half of a degree. Glenn immediately received the report and confirmed that he was a "go for seven orbits." It was a message that was so often misinterpreted by the press to mean that Glenn was actually supposed to have completed seven orbits of the earth. Of course it was nothing more than an advisory that his trajectory would be stable beyond the planned three orbits. During the precise recalculation of the trajectory data, the computer even calculated a stable orbit for almost 100 orbits of the earth.

Although the separation jets released the capsule from the rocket at exactly the right time, the five-second-long damping phase began two and a half seconds too late. This small pause caused a substantial deviation over the roll axis at the precise moment that the turning maneuver began. While the attitude control system easily managed the deviation, it consumed 5.5 pounds of fuel. Glenn looked out the window and could see the Atlas, now 100 feet away, tumbling end over end through space.

The first orbit was textbook. The tracking station on the Canary Islands reported that all onboard systems were in perfect condition. Glenn reported to the Canaries that he was a little behind schedule but that all systems were "go." Over Kano Glenn carried out his first major attitude control maneuver. He confirmed the yaw control and positioned the capsule so that the nose was pointing in the direction of flight. Previously, like Shepard and Grissom, he had always had his back to the direction of flight, so that he could fire the retro-rockets at any time.

Over the Indian Ocean Glenn became the first American to observe a sunset from orbit. Then he flew towards the dark side of the earth. There he completed an observation program in which he attempted to identify landmarks, weather phenomena and stars. Over the Muchea tracking station in Australia Glenn had a relaxed conversation with Gordon Cooper. He was also able to see the lights of Perth and Rockingham and identify stars.

Over Canton Island, Glenn readied the periscope for the first sunrise of his orbital journey. When he looked through the optical instrument he saw literally thousands of tiny points of light floating around his capsule and following him. Glenn's first impression was that he was tumbling or was looking into a field of stars, but a quick glance out the window immediately corrected this illusion. He observed that these glowing "fireflies," as he called them, were coming from above and outside his spacecraft. When Friendship 7 reentered the sunlight over the Pacific the "fireflies" disappeared again.

Apart from the mystery of the fireflies and isolated problems with the communications system, the mission went according to plan. At first the attitude control system functioned perfectly. But

Friendship 7 lifts off on its historic first orbital mission of the American space program.

then one of the yaw thrusters began giving him problems, which, as he later said, "stayed with me for the rest of the mission." That was rather disheartening news for the flight control team, which remembered all too well that a stuck valve had ended Enos' mission prematurely.

Glenn noticed that the space vehicle's automatic stabilization and control system was permitting it to drift to the right at a speed of about 1.5 degrees per second. The capsule behaved like an automobile whose front wheels are misaligned. The drift triggered a signal in the system that should have caused one of the yaw thrusters to fire, but it didn't. Glenn switched back to the proportional manual mode and with it brought Friendship 7 into the proper attitude. Then, trying mode after mode, he attempted to determine how he could maintain position with the minimum consumption of fuel. He finally reported to Mercury Control that the fly-by-wire method was the most effective.

After about twenty minutes, with Glenn over Texas, the defective thruster mysteriously began working normally again. Glenn switched back to automatic mode, but after about a minute the opposite thruster began showing the same symptoms. The astronaut again went through all variations, but in the end realized he would have to live with the problem and that he was going to have to be a full-time pilot for the rest of the mission. In any case he no longer had any faith in the automatic system.

Flight control at the Cape noted that Glenn was coping well with his problem, even though it meant that he had to abandon his own observational tasks. But then a more serious anomaly drew their attention. The engineer in Mission Control responsible for the landing system, a man by the name of William Saunders, noticed that Segment 51, a group of sensors that provided telemetric data about the condition of Mercury's landing system, was transmitting an unusual indication. If the signal was correct, then the heat shield and the compacted pneumatic landing bag were no longer in the secured and locked position. In this case the vital shield

This picture was taken by the automatic camera on board Friendship 7 during the orbital phase of the flight.

would only be held to the capsule by the straps of the retro-rocket pack.

Mercury Control immediately advised all ground tracking stations to keep an eye on Segment 51 and to casually work into radio traffic with the pilot that he should be sure to keep the pneumatic landing bag switch in the "OFF" position. This secretive behavior quickly aroused Glenn's suspicion that something was wrong, however. After tracking station after tracking station posed similar-sounding questions and gave identical instructions, he was able to put two and two together.

Meanwhile, the operations team was pondering how to get the capsule safely through the atmosphere with a loose heat shield. They decided not to jettison the burnt-out retro-rocket pack, as its straps were apparently the only thing still holding the shield in place. A telephone call from mission control to Maxime Faget, the capsule's chief designer, determined the strategy. Faget confirmed that this procedure was the correct one, but only if all the retro-rockets fired. If not, then the package would have to be jettisoned, for otherwise the unburned fuel would explode when the capsule reentered the atmosphere. Other than that, however, it should work. The retaining straps would burn away, but by then the dynamic pressure on the shield would be sufficient to hold it in place.

While all of these deliberations on the ground, of which Glenn at first knew nothing, were going on, Friendship 7 had begun its second orbit. The astronaut carried out his observations as far as time permitted, as he was by then flying his space vehicle manually. Over the Indian Ocean he tacked another problem that had been present since liftoff. He had to adjust the cooling water system in his space suit, for it was becoming, as he later observed somewhat ironically, "comfortably warm." The change led, however, to an increase in humidity inside the capsule, and so he had to find a balance between comfortable temperature and more allowable humidity.

The next warning light appeared over Australia and informed him that remaining fuel had dropped to sixty-two percent. Mercury Control recommended

that he let the capsule fly in free drift for a while with no active attitude control.

During the third orbit, Walter Schirra, who was functioning as CapCom in California, officially advised Glenn to leave the retro pack on the capsule after firing the retro-rockets. This meant several deviations from normal reentry mode. To bypass the automatic systems, Glenn had to retract the periscope by hand and manually initiate the reentry sequence. Four hours and thirty-three minutes after launch, with Friendship 7 nearing the Californian coast during its third orbit, the first retro-rockets fired. Glenn checked in with Schirra and declared, "Boy, that felt like I was flying back to Hawaii." Seconds later the two remaining rockets also fired.

Now came perhaps the most dramatic and critical moment of the entire Mercury program. In the control center, the tracking stations and on recovery ships around the globe, engineers,

The Mercury heat shield and retro-pack: three small thrusters (red) around the capsule to distance the capsule immediately after separation from the launch vehicle, and three large ones for the de-orbit maneuver towards the end of the mission. Three grey tension straps hold the solid-fuel rockets to the base of the capsule.

technicians, doctors, recovery personnel and the other astronauts stood nervously, staring at their consoles and trying to make out something from the static. Was the report about Segment 51 correct? And if so, would the retropack straps keep the shield in position until it was fixed in place by the airstream? Would this mission, America's first orbital flight, end with the astronaut being burned alive? The entire Mercury team felt like the defendant in a trial before the reading of the verdict.

Friendship 7 began its long descent to the earth that led across the entire United States to, hopefully, a splashdown in the Western Atlantic. The tracking station at Corpus Christi advised Glenn to at least leave the retropack on the capsule until the pressure reached 1.5 g. Glenn was meanwhile busy overcoming the problems with his flight control system. He flew the capsule in analogue mode and intended to use the fly-by-wire method as a backup. Then he reached the upper limits of the atmosphere.

Almost immediately, Glenn began hearing noises as if someone was rubbing against the capsule. "It's a real fireball out there," he reported to the Cape with a trace of uneasiness in his voice. Then one of the retropack straps snapped around, fluttered briefly in front of the window and was gone. This was followed by smoke and fragments, when the entire installation succumbed to the friction heat. The control system kept the capsule's attitude stable, but the fuel supply for the manual system quickly dropped to fifteen percent, with the maximum braking deceleration still ahead. At this point Glenn switched to fly-by-wire mode and to the fuel tank for automatic control.

Friendship 7 now came to the fateful point in its journey. The tremendous friction heat completely shrouded the capsule in a flume of ionized gas and Glenn experienced the worst emotional stress of his flight. "I thought that retropack was long gone, and now I again saw large pieces of some kind of material flying past my window," he later said. He feared that it was pieces of his heat shield and that it was beginning to disintegrate.

He also knew that if it was, there was nothing he could do about it.

Soon after he had passed this region, the spacecraft began to oscillate so violently that Glenn could no longer control it manually. Friendship 7 swung far beyond the "tolerable" ten degrees either side of the neutral position. "I felt like a falling leaf," he later remembered. So he engaged the auxiliary damping system, which caused the roll and yaw rate to drop to an acceptable level, but only at the cost of excessive fuel consumption. Most importantly, however, the heat shield remained in place.

Finally, both fuel tanks were completely empty. The tank for the automatic control system yielded its last drops 111 seconds before deployment of the drogue chute, while the tank for the manual control system went dry fifty seconds later. Correction of the capsule's attitude ceased and the oscillations immediately began again. At an altitude of seven miles Glenn decided to manually jettison the drogue chute instead of risking it being in a position in which the nose of the spacecraft was pointing down instead of up.

Mission Data	
Mission Name	MA-6
Callsign	Friendship 7
Date	February 20, 1962
Launch Site	Cape Canaveral, Launch Complex 14
Launch Vehicle	Atlas D (Number 109D)
Spacecraft	Mercury Number 13
Crew	John Glenn
Spacecraft Weight	2,888 lbs
Flight Path	Orbital
Orbits	3
Perigee and apogee	99 x 165 miles
Flight duration	4 hours, 55 minutes, 23 seconds
Inclination	32.5 degrees
Landing site	western Atlantic
Recovery ship	USS Noa

But just as he raised his hand to the switch, the drogue chute shot out on its own. The swinging motion quickly settled down.

For the rest of the flight all systems worked as anticipated. At a height of four miles he extended the periscope again. Glenn also tried to pick out something through the window, but this was not easy for it was dotted with smoke residue and covered in an oily film that had obviously come from the melting heat shield or the reaction control jets. The antenna section was jettisoned and then Glenn saw the main parachute ejected and deploy. The control center reminded Glenn to release the pneumatic landing bag. He flipped the switch, saw the green confirmation light and felt a comforting bump when the shield separated and the pneumatic landing bag deployed. Through the periscope, Glenn saw the water approaching and tensed his muscles. This was followed by the impact, which he found more reassuring than frightening. The capsule bobbed up and down in the water and Glenn began checking leak-tightness and structural integrity.

Friendship 7 had landed about forty miles from the target and there was no recovery helicopter to be seen anywhere. The calculations had not considered the possibility that the spacecraft might consume all of its fuel. The destroyer *Noa*, however, was not far away, as it had been positioned to observe the capsule's descent.

It turned out that the destroyer was only six miles from Friendship 7. Just seventeen minutes later the destroyer was alongside. A sailor who was not in fact trained for the job removed the spacecraft's antenna and bosun's mate David Bell made fast a derrick line to pick up the capsule. While being lifted aboard, the capsule banged hard against the destroyer's side several times.

After Friendship 7 had been lowered onto the deck of the destroyer on a pallet, inside Glenn began removing the front panels with the intention of leaving the capsule through the upper tunnel. But he was slowly becoming too hot and he let it be. By radio he instructed the crew to step away from the hatch, removed the safety pin and hit the

A helicopter picks up John Glenn from the destroyer Noah to deliver him to the aircraft carrier USS Randolph thirty-one miles away.

detonator button with the palm of his hand. The detonator sprang back and scratched Glenn's knuckle through the glove of his space suit, causing the only injury of the entire spaceflight. In any case, the hatch was open. Eager hands pulled the grinning astronaut from the capsule. His first words were, "Man, was it hot in there."

The Mercury recovery team doctors described Glenn as very warm, completely sweat-soaked and not very talkative. After several glasses of water and a shower he became more communicative. Glenn had lost six pounds in that short time. During the flight he had only consumed six cubic inches of water in the form of apple puree. The greater part of the weight he lost was attributed to perspiration in the overheated capsule while waiting on the ship.

John Glenn and his wife Annie, together with Vice President Johnson, during the parade through Cocoa Beach.

While still on board the *Noa*, Glenn received a congratulatory phone call from President Kennedy. He was safe and 100-million Americans ended their watch on radios and televisions. For John Glenn, however, the hustle and bustle was just beginning. There were visits with President Kennedy and Vice President Johnson, a parade in Washington before an estimated 250,000 people, another parade through Cocoa Beach and Cape Canaveral, a speech to both houses of Congress, a John Glenn Day (the

first of March) in New York, and honors in the headquarters of the United Nations and in Glenn's home town of New Concord, a small town of just 2,300 inhabitants, where 75,000 people came out to greet him.

Glenn's spacecraft had also survived the rough trip well. A little discoloration here, a few smoke residue traces there, a smudged window, a few scars and a deep scratch on the heat shield caused by the retropack, that was it. Otherwise the capsule didn't look much different than after missions MA-4 and MA-5. The inspection team also found that the heat shield had not separated at all. The reason for the warning indication was a loose contact on a switch. This led to a false jettison signal that had so unsettled everyone during the flight.

After the detailed analysis Glenn's spacecraft, McDonnell capsule number 13, went on a trip round the world. It was generally acknowledged as Friendship 7's fourth orbit of the earth. All over the earth, millions of people stood in line to get a look inside. Today, the small spacecraft can be seen in the Air and Space Museum of the Smithsonian Institute in Washington, next to the original Wright Brothers machine of 1903 and Charles Lindberg's *Spirit of St. Louis*.

THE SLAYTON CASE

The only really tragic case in Project Mercury did not result in the death of an astronaut. It did, however, lead to a massive personal disappointment for the main participants and caused an enormous public reaction.

The pilot and alternate for Mercury-Atlas 7, Donald Slayton and Walter Schirra, had trained with Glenn and Carpenter since they had been selected for the job by Gilruth. On March 15, 1962, without prior notice, NASA announced that Slayton was going to be replaced by Scott Carpenter because

of cardiac arrhythmia. The suddenness of this announcement surprised everyone, especially the journalists, who were already putting together their human background stories about Slayton. The very obvious question was: after all the entrance examinations, after his past as a test pilot, and especially after the constant medical monitoring during training, how could someone suddenly develop such a condition? For more than two years he had been under constant medical observation, and detailed EKGs had been made during almost every physically demanding exercise during that time. And now, just weeks before his launch, they came with this.

It had been known since 1959, however, that Slayton occasionally had a minor heart murmur. This peculiarity was never noticed during his time as a combat and test pilot. It appeared at intervals of several weeks and then only for a short time, and it never showed up during his routine physicals. Slayton himself knew nothing about it until August 1959, when it was first detected in conjunction with centrifuge training. At the time the astronauts' physician, William Douglas, had consulted the chief cardiologist of the Philadelphia Navy Hospital. The latter confirmed what Douglas already believed: namely that the condition was harmless.

Williams wanted to get a second opinion, however, and so he and Slayton visited the Institute for Aviation Medicine in San Antonio, Texas. There Slayton underwent another thorough examination and received the same information as in Philadelphia. Not until later did Douglas learn that the doctor who had performed the examination in San Antonio had written a letter to NASA administrator James Webb in which he recommended that Slayton not be allowed to fly.

The matter passed through several medical centers. In the autumn of 1959, Douglas informed Mercury director Robert Gilruth, who in turn advised NASA headquarters in Washington. Douglas also passed the information on to the head of the air force medical service and received instructions to proceed no further. The matter appeared to go to sleep, Slayton's health showed no abnormalities, and in November 1961 he was selected to be the pilot for Mercury-Atlas 7.

Shortly after New Year, James Webb remembered the letter from the air force doctor in 1959 and ordered a new investigation of the matter. Several boards with a large number of doctors continued to view Slayton's symptoms as harmless and recommended that he maintain his flight status. The last medical board consisted of eight air force doctors, who examined each facet of the case and of course Slayton himself, and ultimately reached the decision

Deke Slayton.

that Slayton was, from a medical point of view, fully qualified both as an air force pilot and astronaut.

Administrator Webb then gave the case to three national cardiology authorities: Proctor Harvey, professor of cardiology at Georgetown University, heart specialist Thomas Mattingly of the Washington Hospital Center, and Eugene Braunwall of the National Institute for Health. Their report to Webb stated that they were not in a position to definitively say whether Slayton's physiological performance would be impaired by these symptoms. But they did recommend that, "if NASA has an astronaut without these symptoms, they should use him instead of Donald Slayton." Their report cost Donald Slayton his spot in the Mercury-Atlas 7. He had already named it Delta 7.

Donald Slayton was bitterly disappointed by this decision. He remained in the astronaut corps but increasingly took on more administrative tasks and finally was named NASA's first chief astronaut. In this position he was responsible for selecting all crews for the Gemini, Apollo and Skylab projects. He maintained his flight status through all those years, and, in July 1975, finally went into space as part of the Apollo-Soyuz test project.

AURORA 7 LANDS OFF TARGET

One could have assumed that Wally Schirra, Slayton's alternate, would now step in to take his place. But with a view to all of the many delays and numerous additional training modules that Carpenter would have had to go through as Glenn's alternate on MA-6, Walter Williams decided that Carpenter should take over mission MA-7. This was simply because of his advantage in training over Schirra.

Glenn had proved that a man on a spacecraft was definitely more than just a passenger. The planners of the Manned Spaceflight Center modified the mission for Carpenter so that there were more control tasks for him as pilot and more jobs in the role of scientist. Carpenter was to carry out combined roll and yaw maneuvers, in order to be able to watch the sunrise. There would also be maneuvers in which he could use landmarks and the day-night terminator for navigational purposes. He was to test the effectiveness of celestial navigation and fly for long periods with his head towards the earth in order to test the effects of this maneuver on the pilot's orientation ability. Other scientific tests included releasing a colored balloon attached to the capsule by a line, and observing liquids and their surface tension behavior. He was to attempt to spot a flare fired from the ground, take weather photographs with a special camera in various frequency ranges, observe the phenomenon of airglow, and much more.

The balloon was a thirty-inch inflatable Mylar sphere. It was housed deflated in a small container in the antenna canister along with a gas cartridge and about ninety-eight feet of line. The entire system weighed about two pounds. The ball was covered with colored geometric patterns. Carpenter was to release the balloon at the perigee of the second orbit. The purpose of the experiment was to provide atmospheric drag and color visibility data.

Carpenter inspects his space capsule. Here the heat shield is still missing, so that the shock-absorbing honeycomb structure and conduits can be seen.

Based on information from mission MA-6, a series of modifications was undertaken. In particular, equipment was reduced to save weight. The SoFar bombs and the radar reflector were deleted, because the Sarah radio beacon and the water dye markers had proved their effectiveness. Then the knee and hip belts were removed. The red filter in the window was deleted, as was the flight path indicator, one of the heaviest instruments in the instrument panel. One of the two cockpit cameras was also eliminated.

When Carpenter learned in March 1962 that he was going to fly Slayton's mission, he chose the call sign Aurora 7 to indicate the dawn of a new age. The seven, of course, stood for the seven Mercury astronauts. There was also a small background meaning, for as a child in Boulder, Colorado, he had lived at the intersection of Aurora and Seventh Avenue. The capsule with the McDonnell serial number 18 had been at the Cape since November 15, 1961. Its launch vehicle, Atlas number 107D, arrived on March 8, 1962.

Compared to previous Mercury missions, preparations proceeded smoothly. Carpenter's flight would have been possible in April 1961, but major maneuvers by the Atlantic Fleet, in which the recovery vessels also took part, delayed the possible launch date until May 20.

In his quarters on the second floor of Hangar S, Scott Carpenter was awakened at 01:15 in the morning. For breakfast he had filet mignon, poached eggs, orange juice, toast and coffee. This was followed by the usual medical examination and the attachment of the bio-sensors. At 03:25 he began putting on his space suit, assisted, as were all the astronauts, by Joe Schmitt. Even with all that he was ready too early, and so he waited in a reclining chair on the bus that would take him to the launch pad.

At 03:45 Carpenter and his party, including Joe Schmitt and John Glenn, left Hangar S, climbed into the transport and made their way to Launch Complex 14. Still on the bus, the astronaut was briefed on the weather forecast for the Cape, where there was still thick ground fog, and the primary

Carpenter casts a probing glance inside his spacecraft Aurora 7.

and alternate landing zones. At 04:35 he entered the lift. Then he climbed into the capsule and the hatch was bolted shut. This time there were no problems with the seventy bolts.

The countdown went smoothly, but the ground fog was as heavy as before and the high cloud was also quite thick. Just eleven minutes before the planned launch time Mercury Control decided to wait until the fog dissipated and the cloud thinned. This was necessary in order for the cameras to be able to follow the rocket's ascent. At 07:45, after the smoothest countdown of an American manned mission, Atlas 109's engines roared to life.

The booster ran much smoother than Carpenter had expected from the accounts of his colleagues. He found the vibration level after liftoff low, but during the phase of maximum dynamic pressure the noise level inside the capsule was considerable. BECO was quiet and smooth. The sustainer engine ran until T + 5:20 minutes, the vernier engines a few seconds longer.

A double bang indicated that the adapter ring had released the capsule and that the control jets had separated it from the booster. Then it was time to place the space vehicle into the normal orbital position, with the heat shield forward. Glenn had left this maneuver to the automatic control system and consumed six pounds of fuel in the process. Carpenter flew manually and used the fly-by-wire system, which consumed just two pounds of fuel.

The astronaut was able to see his launch vehicle for a long time, even as he passed the Canary Islands. Over Kano, Nigeria, he reported to the tracking station that he was behind schedule because he was having difficulties loading the camera with the special film with which he was supposed to photograph the horizon. That was at about the same time that the temperature in his spacesuit began rising.

On the dark side of the earth he determined that usable star observations were not possible because it was simply too bright inside the capsule. Because of the poor visibility they did not seem to him to be particularly useful for navigation purposes.

The Canton Island tracking station worried about his core temperature, which was at thirty-nine degrees. Carpenter remarked, however, that he was halfway comfortable, although he was sweating. The high temperature levels continued for the rest of the mission and led to the CapComs at the tracking stations regularly inquiring about Carpenter's physical status.

During the second orbit, Carpenter made frequent maneuvers using the fly-by-wire system and the attitude control system's proportional manual mode. He turned the ship around to take photos, he inclined the capsule eighty degrees to see the flashing beacon at Woomera, he carried out yaw maneuvers to observe the airglow phenomena and rolled the capsule for an inverted flight experiment. He also positioned the capsule vertically until the antenna container was pointing straight at the earth and found the sight breathtaking.

Carpenter used his attitude control system extensively to work out his program of experiments and observations. In doing so, six times he inadvertently activated the so-called double authority control, in which both the automatic and redundant systems were in operation at the same time. At the end of the second orbit Carpenter's remaining fuel for manual control was forty-two percent and for automatic forty-five percent. The ground stations began urging him more frequently to conserve fuel.

One hour and thirty-eight minutes after launch, Carpenter released the balloon, but it failed to inflate properly. It remained "sausage-shaped" and the different colors were almost impossible to make out. The balloon's movements were unstable, yet he was able to complete several of the atmospheric drag measurements. Then he began another maneuver with the attitude control system, and the line wrapped around the antenna container. Carpenter wanted to jettison the balloon, but the thing stubbornly remained close to the spacecraft even during the third orbit of the earth.

The fuel situation was slowly becoming critical, and so he had no choice but to go into free drift mode. With each sunrise Carpenter saw the fireflies,

This photo was taken at the instant that Atlas D
Number 107's engine ignited. In the foreground is
the water drainage channel leading away from the
flame shaft.

or the "Glenn effect" as the Soviets called it. Although the glowing particles appeared to move at different speeds, they remained near the spacecraft. At sunrise of the third orbit, Carpenter reached for the densitometer, a device that measured light intensity. As he did so he accidentally struck the hatch with his gloved hand, and suddenly he saw a cloud of glowing particles fly past the window. A second tap on the hatch produced another cloud. This chance happening explained the mystery of the fireflies. The outside of the space vehicle was obviously covered with frost. The slightest tap caused some of the frost to loosen. The mission was very successful until just before retro ignition, apart from high fuel consumption. But as Carpenter had spared fuel during his third orbit, the remaining forty percent should have been easily sufficient for the reentry maneuver. The tracking station on Hawaii instructed Carpenter to begin the pre-retrofire countdown and to switch from manual flight control to the automatic stabilization and control system.

When Carpenter carried out this instruction, he suddenly found himself in difficulties. The automatic stabilization system refused to maintain the thirty-four-degree pitch angle and zero-degree roll angle for reentry. When he tried to find out what was wrong, he fell behind in his reentry checklist. When he was still unable to find the problem, he switched from automatic to fly-by-wire mode, but forgot to switch off the manual system. For ten minutes the space vehicle consumed fuel from both systems simultaneously.

Finally, Carpenter believed he had succeeded in aligning the spaceship for the retro sequence. When the capsule was over Point Arguello, Alan Shepard transmitted the countdown for retro ignition. After the automatic system failed to function properly, Carpenter had to manually operate the ignition button for the retrorockets. It was seconds after Carpenter transmitted "Mark! Fire one!," however, that the first retrorocket fired. The second and third followed in quick succession.

Carpenter had made a significant attitude control error while manually positioning the spacecraft. When he initiated the retro sequence the yaw angle had a deviation of twenty-five degrees to the right of the correct position. As a result, the thrust vector was not aligned with the flight path vector. This anomaly alone would have caused Carpenter to miss the planned splashdown point by 174 miles. The delay in ignition of the retrorockets contributed three seconds, which added about another fifteen miles. Unfortunately, the output of the retrorockets was three percent below nominal value, which led to another fifty-nine mile deviation. Over Point Arguello it was already clear that splashdown was going to take place far from the target area.

Not until the retrorockets had fired did Carpenter notice that the manual control system was still active. He quickly switched off the automatic fly-by-wire system. The indicator still showed six percent fuel, but in fact the tank for the manual system was already empty. Then Carpenter checked the automatic system. It showed a fuel reserve of fifteen percent, but the astronaut wondered if this indication was accurate. With this gnawing doubt and the knowledge that it would be another ten minutes before he reached the uppermost level of the atmosphere, Carpenter kept his hands off the controls. Whatever fuel there was left had to be reserved for the critical turbulent phase of reentry.

Finally, Aurora 7 reached the 0.05 g deceleration point. Soon the capsule began oscillating wildly. Carpenter engaged the damping mode and by radio Gus Grissom reminded him to close his helmet visor.

At an altitude of sixteen miles the fuel tank for the automatic control system also ran dry. The oscillations became heavier as the capsule flew through the ten-mile mark. Aurora 7 swung far outside the tolerable ten-degree limit and Carpenter had tried to release the drogue chute at an altitude of nine miles to stop the wild movements, but he forced himself to hold out against even stronger pendulum movements before firing the launcher for the parachute at a height of five miles. At three miles Carpenter armed the main parachute, which

Mission Data	
Mission Name	MA-7
Callsign	Aurora 7
Date	May 24, 1962
Launch Site	Cape Canaveral, Launch Complex 14
Launch Vehicle	Atlas D (Number 107D)
Spacecraft	Mercury Number 18
Crew	Malcolm S. Carpenter
Spacecraft Weight	2,976 lbs
Flight Path	Orbital
Orbits	3
Perigee and apogee	95 x 161 miles
Flight duration	4 hours, 56 minutes, 5 seconds
Inclination	32.5 degrees
Landing site	western Atlantic
Recovery ship	USS Farragut

Like all of the other astronauts, Carpenter also practiced exiting through the narrow tunnel during training.

was fully deployed by two miles. Soon afterwards he activated the pneumatic landing bag.

Splashdown was noisy but not as rough as the astronaut had expected. The capsule listed sharply to the left and as the moments passed did not right itself. The astronaut noticed several drops of water on his tape recorder and wondered if Aurora 7 was going to share the fate of Liberty Bell 7.

Carpenter was aware that he had overshot the target area by a wide margin. In his last radio transmission, Grissom had advised him that the rescue diver would take at least an hour to reach him. He tried to reach someone by radio, but no one answered. It seemed to him that the capsule was sitting low in the water. It was thus too dangerous to blow the hatch, as the capsule would fill with water. All that was left was the egress method the capsule's designers had originally come up with: up and out through the narrow tunnel. It

was thirty-eight degrees in the cabin and Carpenter was sweating all over. He took off his helmet, removed the forward bulkhead and began working his way up through the narrow tube. It was a hot, exhausting exercise, complicated by the fact that he had with him his camera, the packed inflatable raft, survival equipment and the twisted hose connections of his space suit. Finally he got his head outside.

When he had crawled out of the hatch to chest level, Carpenter braced himself on the edge of the nose cone on his elbows for a moment and removed the hose connection on his suit. He forgot, however, to put the neck dam in the correct position and close the valve on the end of the hose. The sea state was acceptable, the waves about five feet high. Carpenter removed the camera, squeezed out of the capsule nose and let himself slowly slide into the water. The life raft inflated but remained upside down. Carpenter now noticed that the inside of his space suit was getting wet. He realized he had forgotten to close the valve and did so. Then he held on to the capsule with one hand and righted the life raft with the other. He scrambled into the small raft, got his camera and prepared to wait as long as it took until he was found. The radio beacon functioned normally and the marker dye colored the water around him green.

The status of Carpenter and Aurora 7 had not been revealed to the public. Everyone following the flight by radio or television knew that he had to be down by now. But was the astronaut safe? The public did not know that a P2V Neptune had detected his radio beacon's signal from a distance of fifty miles. Another Neptune had located the signal from a distance of 250 miles. This made possible a triangulation that precisely pinpointed Carpenter's location. As well, eight minutes before splashdown an SA-16 flying boat had taken off from Puerto Rico to head for the splashdown point derived from radar data. Three ships were also not that far away: a Coast Guard cutter abeam the island of St. Thomas, a merchant vessel about thirty miles away, and the destroyer *Farragut*, which was seventy-five miles southwest of the capsule's position. In any case, it would be more than an hour before any of the recovery forces reached the splashdown point. As the life raft had no radio, the drama intensified for the public.

Carpenter was now relatively comfortable in his life raft. Thirty-six minutes after splashdown he saw two aircraft approaching: a P2V and—quite unexpectedly—a Piper Apache. The two aircraft began circling over him, and he knew that he had been found. Twenty minutes later he saw two SC-54s arrive. Two frogmen parachuted from one of the aircraft, but Carpenter, concentrating on the other aircraft, failed to notice.

One of the two men was Airman First Class John Heitsch, who had jumped from one of the SC-54s one hour and seven minutes after Carpenter's splashdown. He landed quite a distance from the space capsule, released his parachute and swam over to Carpenter's life raft. "Hey!" the frogman called to the completely surprised Carpenter. Taken aback, the latter asked: "How did you get here?" The second parachutist, Sergeant Ray McClure, arrived soon afterwards. The two frogmen quickly inflated two more rafts and made them fast to the space vehicle. McClure and Heitsch later described the astronaut as cheerful, in a good mood and definitely not exhausted. Carpenter opened his survival pack and offered both a bite to eat. They gratefully declined but drank some of Carpenter's water ration.

Now they were three, though still without radio contact, and they watched the growing number of aircraft circling overhead. One of them dropped a flotation collar for the spacecraft. It struck the water with a loud bang and one of the compressed air pumps for inflating the collar was broken. The two divers fetched the collar and attached it to the capsule, but only the lower of two rings would inflate. Afterwards they crawled back into their life rafts.

Seconds later a parachute and box dropped gently into the water. The frogmen assumed that it was the long-awaited radio set. One of the two swam the considerable distance to collect it. He returned with the box; the pair opened it and inside found a battery but no radio. This moment, with the furious cursing of the two frogmen, took on a particularly anecdotal note in Carpenter's subsequent report.

Ninety minutes after splashdown the Grumman SA-16 from Puerto Rico also arrived in the now crowded airspace over Carpenter's splashdown site. The sea looked calm enough to risk a landing and pick up the astronaut. To Mercury Control at Cape Canaveral, however, it seemed too risky, and because of the stable situation on scene, they wanted to continue with the normal procedure.

Three hours after splashing down, Carpenter was finally picked up by an HSS-2 helicopter. An unfortunate combination of a brief lowering of the winch and a high wave resulted in the astronaut being completely submerged. The especially humorous part was that Carpenter was determined to keep the film dry, and so all that was seen of the submerged astronaut was his raised arm holding the camera. Then the wet adventure was over and Carpenter was initially deposited on the *Farragut*, where he received the now traditional call from President Kennedy. The president expressed his relief that Carpenter was safe, and Carpenter apologized for "not having aimed better during landing."

Four hours and fifteen minutes after splashdown, Carpenter arrived aboard the aircraft carrier *Intrepid*, where the medical examinations began. Aurora 7 was picked up by the destroyer *Pierce* and reached Cape Canaveral the following day. When picked up, Aurora 7 was listing about forty-five degrees instead of the usual fifteen to twenty degrees and there were about fifty-five gallons of water inside. Apart from that the capsule was in good shape, similar to the other missions.

In the post-flight debriefings the astronaut observed that every single task had taken more time than allocated. Some of the equipment was difficult to operate, for example the film that had to be loaded in the camera. The ultimate flight plan had not been available until just before the mission. Carpenter declared that the final flight plan should be available at least two months prior to the flight. With regard to fuel consumption, Carpenter stated that, for the purpose of attitude control, the six engines for coarse adjustment could be shut down, as they were unnecessary for the orbital phase.

This rather blurry photo was taken from an aircraft circling over Carpenter's splashdown point. It shows the situation after the two rescue swimmers reached the capsule.

A TEXTBOOK FLIGHT

Walter Schirra received his mission assignment on July 27, 1962: a six-orbit mission with Mercury-Atlas 8. His ultimate flight plan was set down on August 8, almost sixty days before the start of his mission and thus in the time frame stipulated by Scott Carpenter. Schirra had observed very closely what had ultimately led to the difficulties during Aurora 7's mission and insisted that his flight should be rigorously trimmed of experiments and research tasks. Schirra wanted MA-8 to be a pure engineering flight and prove that missions of a day or even more were possible with the Mercury system.

That was no easy task. Both MA-6 and MA-7, which lasted just four and a half hours, had returned home completely out of fuel. Schirra nevertheless went through every detail with the flight planners and agreed to only a minimum number of experiments and observations that required larger movements with the attitude control system. The majority of the experiments were passive, such as test panels for ablation materials or new types of paint. Schirra put a great deal of thought into the mission call sign and in order to underline the engineering character of the flight he chose Sigma 7.

On August 11, production capsule number 16 was almost ready in Hangar S. Schirra's launch vehicle, Atlas number 113D, had also arrived at the Cape when NASA officials grimly became aware of news from Moscow: the Soviets had launched the five-ton spacecraft Vostok 3. On board was Major Andriyan Nikolayev. His orbit had a perigee of 112 miles and an apogee of 155 miles. Inclination to the equator was sixty-five degrees. The next day the "gap" in the American space program compared to the Soviets became a "gulf," for twenty-four hours after Nikolayev was placed in orbit they launched Vostok 4 carrying Lieutenant Colonel Pavel Popovich. Soon after the launch, Nikolayev announced that he had spotted Popovich. At times the two spacecraft were just three miles from one another. Speculation that the two vehicles would attempt to dock proved unfounded, however.

Nevertheless, the Americans were shocked again: Nikolayev landed on August 15 after ninety-five hours in space, during which he circled the earth sixty-four times. Popovich landed six minutes after him after forty-eight orbits of the earth and seventy hours in space. These were values that the USA would not achieve for years.

Mission planning for MA-8 had begun in February, immediately after Glenn's MA-6 flight. A flight with six or seven orbits seemed a logical intermediate step toward the final goal of the Mercury program, a one-day mission. This required a series of modifications to the capsule. Previous flight management had required constant use of the automatic stabilization and control system, which consumed a great deal of fuel and electrical current. The strategy for the oxygen reserves had to be rethought and the tracking and communications network, previously optimized for a three-orbit mission, had to be reconfigured.

Walter "Wally" Schirra piloted Mercury Atlas 8.

The energy needs for a three-orbit mission amounted to 7,080 watt-hours. The total available on board was 13,500 watt-hours. A seven-orbit mission would consume 11,190 watt-hours, which would leave a reserve of less than seven percent and that was too little. The logical consequence was that several power consumers would have to be shut down. An oxygen supply of nine pounds would be needed for the seven-orbit mission. The Mercury, however, was only capable of accommodating two four-pound oxygen tanks. With the lengthening of the mission duration the need to purify the air of carbon dioxide also rose. To date there had been a canister of lithium-hydroxide on board for this, with a capacity of five pounds. That was also too little. The most difficult problem, however, was the one with the fuel. The entire attitude control philosophy had to be rethought in order for the astronaut to have a reserve of one-third of his initial value.

The problem reduced itself a little when, while calculating the necessary recovery forces, it was found that additional ships would be needed for a seven-orbit mission in order to cover all possible emergency scenarios.

It was therefore decided to make just six orbits of the earth. The landing zone was moved from the Atlantic to the Pacific, for there the paths of the fifth and sixth orbits crossed about 274 miles northeast of Midway Island.

The tracking network was bolstered by five airborne relay stations in the form of air force C-130 aircraft. They were to cover areas which otherwise would have been beyond the communications range of the ground stations. In the end the recovery fleet consisted of nineteen ships in the Atlantic and nine in the Pacific. There were also 143 aircraft of various types stationed in the primary and secondary splashdown areas. All in all, no less than 17,000 people took part in the recovery action for MA-8.

The launch vehicle had been delivered to the Cape on August 8, but the air force advised that the turbo pumps had failed on four recent military

Several weeks before launch, Sigma 7 undergoes preparations for mission MA-8 inside Hangar S.

missions by the Atlas. To be sure that this problem did not arise on number 113D, a static test run was scheduled. But before it could be carried out, the air force and Convair inspectors found a crack in a weld seam. Additional tests and the repairs delayed the planned launch date until October 3.

That day Schirra was awakened at 0140 and went through the now standard routine before a mission. The astronaut's breakfast, consisting of eggs, filet mignon, toast, orange juice and coffee, was bolstered by a piece of bluefish, which he had caught the day before. Flight doctor Minners examined the astronaut again and attached the bio-sensors to his body. Then it was the turn of suit technician Joe Schmitt, who helped Schirra into his space suit, as he had previously done for all his colleagues. Shortly after 0400 the astronaut and his entourage left Hangar S and headed for the launch pad. At 0441 he climbed into the space capsule.

The technicians then closed the hatch and each bolt sank smoothly into its threads. The countdown progressed without incident until 0615. Then the tracking station on the Canary Islands reported a problem with its radar equipment. As this station was of great importance in determining the orbit parameters, Williams ordered the countdown stopped. The station required fifteen minutes to correct the problem. The remaining forty-five minutes of the countdown proceeded without further interruptions.

The engines of the Atlas roared to life at 0715 and the booster left the launch pad. Ten seconds above the pad, the roll sensors inside 113D reported an unexpected clockwise rotation. It was only twenty percent below a roll rate that would initiate the launch abort system. Just as suddenly as the movement had begun, it stopped again.

Burnout of the booster engines happened two seconds earlier than programmed. Seconds later the escape tower separated from the capsule, leaving smudges on the window. Acceleration of the cruise engine seemed very slow to Schirra. There was no need to worry, for the sustainer shut down ten seconds later than planned and gave him an apogee that was higher than any attained by an American astronaut: 176 miles. His burnout speed was also the highest ever achieved during the Mercury project: 17,560 miles per hour.

When Sigma 7 separated from the Atlas, Schirra switched the attitude control system to fly-by-wire mode and initiated a leisurely rotary movement, just four degrees per second, in order to achieve the correct position in space. The movement was so slow that, in the end, he had used less than 0.5 pounds of fuel. On Carpenter's recommendation, the fly-by-wire system had been modified to use only the small jets to change attitude. It also enabled Schirra to maintain sight of his Atlas booster for quite a while.

Over Africa it began to get warm inside his space suit. He decided to devote his full attention to the problem. He had witnessed the excitement in the control center during previous flights and wanted to prevent, as he put it, "the people in Mission Control scampering about from excitement"

when they saw the readouts from the temperature sensors. He was not wrong, for mission controller Frank Samonski, who monitored the life support system, had already noticed the steady rise in temperature and consideration was already being given to aborting the mission after the first orbit. Samonski discussed the situation with Flight Surgeon Charles A. Berry. He was of the opinion that Schirra was in good condition and recommended waiting to see if a solution to the temperature problem was found. After struggling with communications problems, Schirra first learned over the Guaymas that he had the green light for further orbits.

Glenn and Carpenter had already encountered the temperature problem, and Schirra had therefore busied himself with it before the mission. When the problem became acute for him too, the temperature switch was in position four of ten possible positions. He knew that changing the temperature too quickly would cause the heat exchanger to ice up. That would make the problem worse instead of better. And so he moved the knob half a mark at a time, waiting ten minutes to see if there was a change and then turning the knob another half mark. From Position 7 it was clearly cooler; at position 8 the temperature was normal. It still tended to be too warm in the space vehicle. If sunlight came through the window the temperature rose immediately. Schirra soon noticed a crust of salt around his mouth and described the situation to ground control like "mowing the lawn in Houston (where temperatures of forty degrees are not uncommon) on a summer day."

Like Glenn and Carpenter, he found the periscope almost useless. Thought had been given to removing the heavy piece of equipment prior to the launch, but they wanted to test it for navigation purposes once again.

Over the Pacific during the first orbit, Schirra went into "chimp mode," meaning that he let go of the controls and floated over the ocean in free drift. By the end of the first orbit he had used just 1.3 pounds of fuel, an entire magnitude less than his two predecessors.

The launch of mission Mercury Atlas 8.

Mission Control at Cape Canaveral during the Sigma 7 mission.

He also saw the fireflies that Glenn had first reported and tapped on the hatch to produce the same effect as Carpenter. Initially he seemed to enjoy the constant check-ins with the ground stations, but as the flight went on he more frequently found it a tiresome task, especially when he wanted to concentrate on his work.

During the second orbit of the earth, on the night side of the planet, Schirra busied himself with the precision of yaw movements in the darkness. He tried it with the stars as reference points and found the moon well suited for such maneuvers. It worked best, however, with the airglow phenomenon on the earth's horizon. He continued flying long stretches in free drift and with the third orbit began experimenting with energy-saving measures, during which he even shut down his attitude control gyroscope.

Halfway through the fourth orbit his helmet visor began fogging up from the inside. Although the problem continued for the next two hours, Schirra initially did not want to clean the visor from the inside, because he feared if he opened the visor his suit's air conditioning might get out of balance again.

During the fifth orbit radio communications became less frequent, as he overflew fewer of the tracking stations that had been positioned for a three-orbit mission. Sigma 7 functioned almost perfectly. Alan Shepard, who was on the Pacific command ship, advised him that the reserves in both tanks were still more than eighty percent. The suit temperature was by now a comfortable twenty degrees. At the start of the sixth orbit Schirra began preparing for his return to earth. He switched from automatic control to fly-by-wire mode and closed his helmet visor, which he had opened in the last seconds, only to find, however, that it immediately fogged up again. He opened the visor again and wiped it clear. The rectifier temperature was good, battery voltage was correct, oxygen pressure optimal. Although it was a little cool for him, Schirra switched the suit temperature to position 8 in anticipation of reentry.

When he came within range of the command ship in the Pacific again, he rechecked his fuel levels: seventy-eight percent in both tanks. Shepard asked him how far he was into the checklist. He had reached the point where the igniters for the retrorockets were charged. Shepard transmitted the countdown for activation. Then came the retrorocket sequence. The first charge fired eight hours and fifty-two minutes after Sigma 7 had lifted off from Launch Pad 14 at Cape Canaveral. Schirra was pleasantly surprised that the automatic attitude control system held the capsule rock-steady during firing of the retros. During the burn phase he looked out the window and the field of stars did not even shake before his eyes. When the burn phase had ended he again looked at the fuel contents gauge and saw the needle swinging between fifty-two and fifty-three percent.

Schirra went back to his preferred fly-by-wire low thrust mode. Then he pushed the button to jettison the retropack and the spent unit spun away. He then switched on the fuel-intensive rate stabilization control system in order to dampen the capsule's oscillations on entering the earth's atmosphere.

Schirra could not hear the hissing noise that Glenn and Carpenter had reported. He was, however, concentrating on the rate control system, because he could scarcely believe what he was seeing. So far he had religiously conserved fuel, and now he saw that fuel was being consumed as if flushed down a toilet. He resisted the desire to switch to a more economical control method, for the engineers wanted to evaluate this system once and for all.

Soon the barometric altimeter began operating. Schirra waited calmly until the needle had reached eight miles. Then he pushed the button for the drogue chute, felt the tension and the parachute opening. Unlike Glenn and Carpenter, who had moved through this zone with no attitude control fuel left and whose capsules had swung wildly, Schirra's descent so far had been very smooth. But whereas the other two had found the phase beneath the drogue chute very quiet—compared to the wild ride before—it now seemed to Schirra as if he had changed from a paved road to a pothole-filled country lane. The attitude control system was still operating with the chute deployed, and its residue began smudging Schirra's window more and more.

Finally the window was literally plastered over and Schirra pressed the fuel jettison button to pump overboard the remaining hydrogen peroxide.

At the three-mile mark he released the main parachute, saw that it was initially reefed and finally deployed fully at a height of two miles. Sigma 7 made what was then the most accurate recovery of the Mercury program. Schirra came down just five miles from the planned landing point, well within sight of the aircraft carrier *Kearsarge* and the cameramen on its deck. Mission MA-8 had ended after a flying time of nine hours and thirteen minutes.

Sigma 7 came down in relatively calm water and sank a little before finally returning swaying to the surface. The astronaut waited patiently about forty-five seconds, he then jettisoned the main parachute and switched on the recovery systems. The spacecraft remained dry inside and the temperature was comfortable. Through the window he could see the green dye coloring the surrounding water.

The recovery helicopters had lifted off from the *Kearsarge* even before Schirra had reached the uppermost layer of the earth's atmosphere. They were on scene immediately. Seconds after the capsule landed in the water, teams of swimmers leapt from the low-flying helicopters and attached the floating collar to the capsule. Schirra reported that he would stay in the capsule and wanted to be towed to the *Kearsarge*, where the crane would lift him onto the deck. Five men then came over in a jolly boat from the aircraft carrier, which was only one-half mile away, and attached a line to the capsule.

Nine hours and fifty-four minutes after launch, Sigma 7 was on the deck. Schirra activated the control for the explosively actuated hatch and sustained the same minor injury to his hand as Glenn. He crawled from the capsule, greeted by cheers and applause from the ship's crew. Schirra looked exhausted but happy. For the next three days the *Kearsarge* was the astronaut's home while he underwent extensive medical testing. First, however, he received the now traditional phone calls from President Kennedy and Vice President Johnson.

Examination of the astronaut and his spacecraft provided outstanding results. Walter Williams spoke of a "textbook flight." Schirra's economical handling of the fuel, in particular, made planning for the crowning Mercury mission considerably easier.

Mission Data	
Mission Name	MA-8
Callsign	Sigma 7
Date	October 3, 1962
Launch Site	Cape Canaveral, Launch Complex 14
Launch Vehicle	Atlas D (Number 113D)
Spacecraft	Mercury Number 16
Crew	Walter M. Schirra
Spacecraft Weight	2,998 lbs
Flight Path	Orbital
Orbits	6
Perigee and apogee	100 x 176 miles
Flight duration	9 hours, 13 minutes, 11 seconds
Inclination	32.55 degrees
Landing site	central Pacific
Recovery ship	USS Kearsarge

FAITH VALIDATED

In fact, the mission of Sigma 7 had gone so well that many senior NASA officials let it be known that they thought that Project Mercury should be brought to an end then and there. Another demanding flight with this first-generation hardware would perhaps stretch the luck that they had enjoyed so far.

More than anything, however, this was interference from the field of politics. There was no doubt among the managers of the Manned Spaceflight Center that the originally planned eighteen-orbit, twenty-seven-hour mission should also be carried out. They thought it necessary if for no other reason than to fill the gap until the first manned Gemini flight, which at the time of Schirra's flight was still scheduled for 1964. It was assumed that the experience gained from a one-day flight would allow the design and layout of Gemini to be accomplished more easily and quickly.

In the summer of 1962, the Mercury program still had four available capsules that had never been flown. They were production numbers 12, 15, 17 and 20. These capsules now received the necessary design changes for a one-day flight. Production number 20 was finally selected for the mission.

On November 9, 1962, the heads of the Manned Spaceflight Center decided to extend the duration of the flight to thirty-four hours and complete twenty-two earth orbits. Of course MA-9 would not be able to match Nikolayev's sixty-four orbits, or the forty-eight of Popovich. A thirty-four-hour mission would, however, at least surpass German Titov's seventeen-orbit flight in Vostok 2. Gordon Cooper was named pilot of the mission, and his alternate was Alan Shepard.

Cooper's capsule was so heavily modified that some spoke of a new design. It received 183 modifications compared to Schirra's spacecraft. Atlas D number 130 was chosen to be the launch vehicle. The biggest problem with all these changes was the tremendous increase in weight. The flight of the MA-9 required an additional oxygen tank, additional liquid coolant, drinking water for the astronaut, sixteen pounds more fuel, and much more equipment, including a 70mm Hasselblad television camera. To at least partly compensate for this added weight, the unpopular periscope was removed. This alone meant a weight saving of almost seventy-seven pounds. The orbital weight of capsule number 20 including pilot finally totaled 3,032 pounds. That was the absolute weight limit for the basic version of the Atlas D.

On the morning of May 14, 1963, twenty-eight ships, 171 aircraft and 18,000 persons spread around the entire globe waited for the launch of Mercury Atlas 9. The mission call sign, chosen by Cooper long before, was Faith 7. The name was a symbol, as he put it, "of my faith in God, in my country and in my fellow team members."

The astronaut was sealed into his spacecraft at 0636 and launch was scheduled for 0900. While Cooper was waiting for the launch, he heard that the secondary control center in Bermuda was having difficulties with its C-band radar. And so he took a nap while repairs were under way. By 0800 Bermuda had resolved the problem, but then there was a problem with the diesel locomotive that was to pull back the gantry. Two more than irritating hours passed before the countdown could continue. But then the radar in Bermuda failed again and finally the launch was cancelled.

When he climbed back into the van, Cooper said, "That was a very realistic simulation." He took off his space suit and went fishing, while at the launch pad the technicians examined the locomotive's fuel pump. The next morning the countdown proceeded without the slightest interruption.

At thirteen seconds after eight, Mercury Atlas 9 began its journey. Cooper's words to Wally Schirra, his predecessor, who now watched over his launch as CapCom, were: "That felt good, boy. All systems are go." Five and a half minutes later he was in orbit. Like Schirra before him, the astronaut watched his booster for several minutes as it spun in the

As this portrait of Gordon Cooper shows, even in the early 1960s NASA knew how to attract publicity.

flight path behind him and then turned to his assignments. Immediately afterwards he completed status checks with the Canary Islands and with Kano in Nigeria. Everything happened so quickly that Cooper could hardly believe that he had already passed halfway across Africa.

During the first two orbits all systems on board the spacecraft worked perfectly. Cooper had just one problem. Once again the space suit's temperature control was functioning erratically. Cooper also complained about an oily film, obviously residue from the escape tower separation, on the outside of his window.

At the beginning of the third orbit he began working on his total of nine experiments. His first task, three hours and twenty-five minutes after the start of the mission, was to eject a six-inch-diameter sphere with xenon strobe lights. At first he couldn't see the small sub-satellite, and it wasn't until the fourth orbit that it entered his field of view. He also saw the flashing light during the fifth and sixth orbits of the earth.

For the second time in two weeks, on May 15, 1963 Gordon Cooper makes his way to the launch pad.

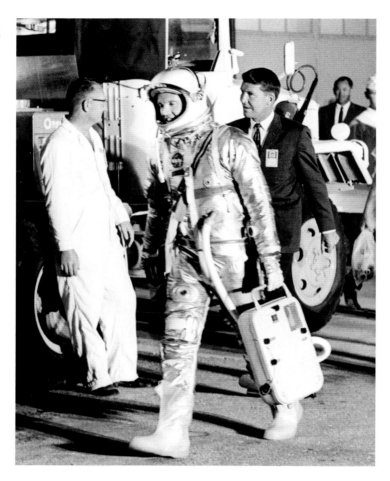

The central tasks of the flight were space-medicine experiments with himself as guinea pig. He ate various types of experimental astronaut food, collected urine samples, took his temperature and measured his blood pressure. During the sixth orbit, Cooper tried to release a balloon from the antenna compartment. It was identical in construction to that of MA-7 and the purpose of this experiment was also to make visual experiments and drag tests. Cooper tried to eject the balloon, but nothing happened. He tried again, and once again nothing happened. As the antenna canister was jettisoned when the parachutes were released during the splashdown phase, they never discovered whether the balloon had ever left the spacecraft or not. When Cooper surpassed Schirra's record at the start of the seventh orbit, he was busy with radiation measurements and experiments with liquid transfer in tubes. Ten hours after launch the ground station in Zanzibar advised him that he had a go for seventeen orbits. Everything was going like clockwork. MA-9 was circling the earth every eighty-eight minutes and forty-five seconds at an angle of 32.55 degrees to the equator.

A rest period was scheduled between the ninth and thirteenth orbits. At the end of the eighth

The last Mercury is on its way. Leroy Gordon Cooper took off in capsule Faith 7, the twentieth and last production capsule made by McDonnell. It was launched into orbit atop the Atlas D with the serial number 130.

orbit he ate and drank a little, shut down most of the systems and placed the spacecraft in free drift mode. But Cooper was still too excited and fascinated to be able to sleep and so he checked in with John Glenn, who was on the tracking vessel *Coastal Sentry* in the Pacific. By then the astronaut had been in orbit for thirteen hours and thirty-four minutes.

Glenn reminded him that it was time to, "tell everyone to go away and leave you alone now."

Cooper relaxed and fell asleep, but he awoke after about an hour because the temperature in his space suit was rising. For the next six hours he slept with interruptions, during which he either took photos of the earth or busied himself with his spacesuit's bothersome heat exchanger. When Cooper arrived over Muchea again during the

fourteenth orbit, he checked all systems thoroughly. The results were all still within the normal range. Ground control was especially interested in the status of his fuel reserves. They stood at sixty-nine percent in the automatic system's tank and ninety-five percent in the tank for the manual system.

Cooper spent orbit fifteen calibrating the equipment and handling messages to and from the earth. The president of El Salvador sent Cooper greetings and Cooper sent a message to a meeting of African politicians in Ethiopia. During the sixteenth orbit he made observations of the northern lights and the airglow phenomenon. During orbits seventeen and eighteen he took infrared weather photos and pictures of the moon as it seemed to sink beyond the horizon into the earth's atmosphere. In between he took photographs of the earth.

The aircraft carrier USS Kearsarge waits for Cooper.

A boat from the Kearsarge takes Faith 7 under tow.

Several ground stations heard him singing as he approached the thirtieth hour of his flight. All the while he had to constantly readjust his suit temperature. He had to eat and drink over the rim of his helmet, which with his bulky gloves was not easy. Then, during the nineteenth orbit, just as he was engaged with a test of the high-frequency antenna over Australia, he noticed the first serious system anomaly of the mission.

Faith 7's 0.05 g warning light came on. Cooper thought that the instrument must be in error, for the articles floating inside the cabin, like the camera he had just been working with, remained in their places. The ground station in California confirmed that Faith 7's flight status was the same as before. Nevertheless, they viewed the indication with concern, for the situation could have effects on attitude control when the retros were fired.

These concerns were justified, for during the next orbit Cooper lost all attitude readings. Then, during the twenty-first orbit, a short circuit occurred

in a bus bar serving the 250-volt main inverter, leaving the automatic stabilization and control system without electric power. The most serious technical failure of all the Mercury flights had appeared during the final orbit of the final mission.

Mercury Control was now in a flurry of worried activity. Faith 7's problems and Cooper's diagnostic measures were crosschecked with identical equipment at the Cape and at McDonnell in St. Louis. Over every ground station that Cooper flew, questions were asked and instructions relayed. The situation was serious, and Cooper decided to take an amphetamine tablet in order to stay focused.

As Cooper passed over the tracking vessel *Coastal Sentry* during his twenty-first orbit, John Glenn gave Cooper a revised reentry checklist. That was at about the time that the carbon dioxide level in the cabin and Cooper's space suit began rising. To Scott Carpenter at the Cape he radioed: "Things are beginning to stack up a little." Twenty minutes later he overflew the Zanzibar ground station and advised that he was going to carry out the reentry maneuver manually.

Twenty-three minutes later Cooper was again in contact with John Glenn. He advised him that he was in retro attitude, holding manually and checklist complete. Glenn gave him a ten-second countdown and Cooper, maintaining the thirty-four-degree inclination manually, fired the retro-rockets on Glenn's "mark" call. Then Glenn said goodbye with the words, "It's been a real fine flight Gordo. Real beautiful all the way. Have a cool reentry, will you." And Cooper replied, "Roger, John. Thank you."

All of the complicated, crowded events of the next fifteen minutes went precisely according to plan as Faith 7 plunged to earth in an ionized stream of superheated gases. Beneath its parachute, the capsule emerged from the clouds four miles from the recovery ship, the aircraft carrier *Kearsarge*, and landed in the gentle waves of the blue Pacific.

Splashdown occurred thirty-four hours and twenty minutes after liftoff. At first the spacecraft floundered in the water a little and then righted itself. Cooper, an air force man, contacted the carrier and requested permission to come aboard. The formal request was answered equally formally. Forty hot and humid minutes after splashdown, technician John Graham, assigned to the aircraft carrier from the Manned Spaceflight Center, blew the explosively actuated hatch from outside. Cooper remained in his couch for eight minutes while a doctor examined him. Then the physician helped him climb out. He was overcome by dizziness for a moment, then he got his balance, and the one-man crew of the one-and-a-half-day Mercury mission walked away in triumph.

Like Schirra before him, Cooper was a little dehydrated. He had lost eight pounds, which he quickly regained after drinking a large quantity of liquids. He had shown that a man was capable of rescuing a mission after vital onboard equipment had failed.

Mission Data	
Mission Name	MA-9
Callsign	Faith 7
Date	May 15-16, 1963
Launch Site	Cape Canaveral, Launch Complex 14
Launch Vehicle	Atlas D (Number 130D)
Spacecraft	Mercury Number 20
Crew	L. Gordon Cooper
Spacecraft Weight	3,031 lbs
Flight Path	Orbital
Orbits	22
Perigee and apogee	100 x 166 miles
Flight duration	1 day, 10 hours, 19 minutes, 49 seconds
Inclination	32.5 degrees
Landing site	central Pacific
Recovery ship	USS Kearsarge

Overheated but happy, Gordon Cooper leaves the spaceship after his historic flight.

EPILOGUE

The heads of NASA met in Washington on June 6-7, 1963, to discuss whether another Mercury mission should be flown. This flight had been in planning for some time, and it was to cover forty-eight orbits and last three days. The necessary equipment was on hand. NASA still had three Mercury capsules and had not taken delivery of all the Atlas rockets it had ordered.

The astronauts, especially Alan Shepard, who was supposed to fly it, were all strongly in favor of this possible mission MA-10. President Kennedy left the decision to NASA administrator James Webb. And he decided against it.

On June 12, 1963, before the senate committee on space flight, he declared that there would not be another mission. Project Mercury was over. Twenty-two months would pass before Virgil Grissom and John Young made the next American manned spaceflight, in a spacecraft that was named Gemini instead of Mercury.

CHRONOLOGY

12/17/1958: NASA Administrator Keith Glennan officially announces the start of Project Mercury.

08/21/1959: The launch of Little Joe 1 with a boilerplate capsule fails.

09/09/1959: The Big Joe mission with a boilerplate capsule is a success for NASA, but a failure for the air force.

10/04/1959: Little Joe 6 successfully carries a boilerplate capsule to an altitude of almost thirty-seven miles.

11/04/1959: Little Joe 1A launches with a boilerplate capsule on a mission to test the rescue system. It is a partial success.

12/04/1959: Little Joe 2 launches with a boilerplate capsule carrying the rhesus monkey Sam and achieves an altitude of fifty-three miles. Mission successful.

01/21/1960: Little Joe 1b launches with a boilerplate capsule carrying the rhesus monkey Miss Sam on a successful test of the rescue system.

05/09/1960: In the so-called "Beach Abort," Mercury production capsule Number 1 is used to simulate a mission abort on the launch ramp. The brief mission is successful.

06/29/1960: Mercury capsule Number 4 is supposed to test reentry into the earth's atmosphere, but the mission fails after a few seconds.

11/08/1960: Little Joe 5 is supposed to test a flight abort under unfavorable conditions using Mercury capsule Number 3. The mission fails.

11/21/1960: Mercury-Redstone 1 is supposed to qualify the Mercury-Redstone combination utilizing capsule Number 2. The mission is a failure but the capsule is undamaged.

12/19/1960: Mercury-Redstone 1A and capsule Number 2 repeat the qualification mission. The Redstone

over-performs, however, and the capsule overshoots the target.

01/31/1961: Mercury-Redstone 2 launches with the chimpanzee Ham on board. He survives the flight intact, but the mission is only a partial success because the launch vehicle again produces too much power and problems with the capsule also develop.

02/21/1961: Mercury-Atlas 2 carries out a suborbital flight with capsule Number 6. The mission is successful.

03/18/1961: Little Joe 5A is supposed to repeat the failed mission of 11/08 with Mercury capsule Number 14. Once again, however, the flight is only partly successful.

03/24/1961: The Mercury Redstone Booster Development Mission, which is to test the changes made after the launch vehicle failures of 12/19 and 01/31, is a complete success.

04/25/1961: Mercury-Atlas 3 is supposed to test Mercury capsule Number 8 in orbit but ends in failure after just forty seconds. The capsule and escape tower are recovered, however, and used again.

04/28/1961: Little Joe 5B is supposed to repeat the failed missions of 11/08 and 03/18 with Mercury capsule Number 14A. Once again, however, the launch vehicle is only partly successful, nevertheless the qualification succeeds.

05/05/1961: First manned flight of the Mercury Program. Alan Shepard completes a successful suborbital mission with Mercury-Redstone 3 and capsule Number 7 that lasts fifteen minutes and thirty-seven seconds.

06/21/1961: Second manned flight of the Mercury Program. Virgil Grissom flies a successful suborbital mission with Mercury-Redstone 3 and capsule Number 11 that lasts fifteen minutes and twenty-eight seconds, however,

the spacecraft is lost during recovery after the escape hatch blows prematurely.

09/13/1961: Mercury-Atlas 4 tests Mercury capsule Number 8A in orbit. Successful repeat of the failed flight of 04/25.

11/01/1961: Mercury-Scout 1 is supposed to carry out a test of the NASA tracking and data relay network. The mission is a failure.

11/29/1961: Mercury-Atlas 5 and Mercury capsule Number 9 carry out the decisive orbit qualification for the first manned Atlas mission. On board is the chimpanzee Enos. The mission is a success.

02/20/1962: The third manned flight of the program. John Glenn completes the first American manned orbital flight with the Mercury-Atlas 6 and capsule Number 13. Duration: four hours and fifty-five minutes.

05/24/1962: Fourth manned flight of the program. Scott Carpenter successfully flies the second American orbital mission with Mercury-Atlas 7 and capsule Number 18. Duration: four hours, fifty-six minutes.

10/03/1962: Fifth manned flight of the program. Walter Schirra completes the third American orbital flight with Mercury-Atlas 8 and capsule Number 16. Duration: nine hours and thirteen minutes.

05/15-16/1963: Sixth and last manned flight of the program. Gordon Cooper successfully completes the fourth American orbital flight with Mercury-Atlas 9 and capsule Number 20. Duration: thirty-four hours and twenty minutes.

06/12/1963: NASA Administrator James Webb officially ends the Mercury Program.